WHO WE ARE?

2021

PHILOSOPHY

5/2021

WHO WE ARE?

Abdenal Carvalho

Copyright 2021 ® Abdenal Carvalho

Title: Who are we? — About the Mysteries of Our Existence

Published Date: 05/2021

Author review

Author's Cover Designer

Category: Self-help / 120 pages / 2nd Edition

Author's translation

This work follows the rules of the New Spelling of the Portuguese Language. All rights reserved.

The storage and / or reproduction of any part of this work is prohibited, by any means — tangible or intangible — without written consent by the author. Violation of copyright is a crime established by law No. 9,610 / 98 and punished by article 184 of the Brazilian penal code.

SUMÁRIO

FOREWORD BY THE AUTHOR...7

FIRST PART 9

01- WE ARE WHAT WE THINK 11

02 - WE ARE RESPONSIBLE 29

SECOND PART 59

03 - WE ARE HUMANLY COMMITTED 61

04 - ARE WE UNIQUE? 71

05-ONLY EVOLUTIONARY BEINGS... 75

06. WHO ARE WE REALLY? 83

07. WE ARE TIME TRAVELERS.. 89

08. WE ARE PERSISTENT... 103

FINAL CONSIDERATIONS... 113

BIBLIOGRAPHY 115

FOREWORD BY THE AUTHOR

Who are we really? Where did we come from and where are we going after death? Did we come into this world by chance or is there a specific reason? Who sent us to this planet and what is our true mission as a person? Philosophy, since ancient times, has been trying to clarify these doubts about the origin of humanity, the real reasons for its existence on this planet and what is the foundation of man being born, growing, dying and where to go after that.

For the most skeptical about a divine being who created us out of the dust of the earth, understanding the mysteries of creation and knowing the reason for being here is fundamental to understand who we really are and what we should do in the contribution of the evolution of our species .

Religions conform to the biblical account that we are creatures of a sovereign God and that we receive the gift of life from him, but science insists on evolutionary theory, stating that we come from monkeys and chimpanzees. The truth is that man brings with him an uncontrollable curiosity about the unknown, what he is unable to understand or explain logically.

Thus, this work tries to fill the mental gaps of my readers in relation to this question focused on what we can really become and why we are in this existential plane. I hope that each of my readers will identify with some of the ideas launched in this book. Good reading.

FIRST PART

01- WE ARE WHAT WE THINK

We are still part of a group formed by thousands of people around the world who ask themselves daily what the true meaning of life is. With the exception, of course, of the most fanatical religious who idealize the concept that there is a divine plan that directs everything in our lives and whose mission of humanity is to be born, to grow, to give good testimonies as a just, perfect and holy being or someone completely imperfect, merciless and dirty by the evil that comes from the darkness and, finally, disincarnate from your physical body, going directly to the light or to the eternal darkness.

What is unable to remain silent within curious minds as to the real reason for our existence in this chaotic and injustice-like world like mine and that of thousands of other individuals who also think the question, which says: Who are we? Where did we come from and where will we go after we leave for the afterlife? Who truly created us, formed us and then sent us to this planet? Does the Holy Bible speak the truth about the creation of human life and all the other things that coexist here?

Or is it science that is in fact right to say that everything came into existence after a cosmic explosion that happened in the Universe millions of years ago? Reading what the Holy Scriptures tell us in Genesis 1: 1 onwards, I wonder if in fact Moses had such a vision of Creation or everything was just reports from extremely creative minds that after writing such information attributed them to Israel's greatest human leader.

It is not my goal here to question the veracity of biblical writings or to argue about any form of religiosity, but to try to understand how we really were created and what we actually came to do in this world where we can clearly see the rise of some and the degradation of others in an unfair plane of existence where only a small part of the people seem to have been chosen to accomplish great deeds and conquests, while the majority remain with their heads buried in the sand of poverty and misery.

If we were really created by a divine being, why the enormous inequality of values, be they material or spiritual? If in fact the cause of this misfortune in humanity is the sin that comes from disobedience that occurred hundreds of years ago in the Garden of Eden and Christ made his salt sacrifice on Calvary to remedy the mistakes of our first parents and to give man a chance to to reconcile with their God, because many of those who already confess Jesus as their only and legitimate Savior still suffer the same injustices and needs?

Is it correct to say that this immensely powerful, rich, merciful and just God wants his children to live in these conditions? The same sacred book proposes that the Creator of all things desires the good of humanity, however, the poverty and misery in which most of it still remains is due to the insistence of living in complete rebellion against the dictates of his holy will. But, were we really created just to satisfy you no more than that? Particularly most of us disagree.

Therefore, what remains for us is to consider that we came from somewhere and here the theory of Evolution proposed by Darwin that proposes the evolution of species comes into play. In the article published by Lohana Ribeiro on 01/02/2019 on the website Educa Brasil, we read that: "According to the Theory of Evolution, individuals undergo necessary changes for their survival. In order to adapt to the environment in which they live.

Who we are? — About the Mysteries of Our Existence

And the first to propose an evolutionary theory was the French biologist Jean-Baptiste de Lamarck, who in 1809 published his theories in the book "Philosophy Zoological". However, the best known person when talking about the Theory of Evolution is the British naturalist Charles Robert Darwin. Author of the book "Origin of Species", created in 1850, Darwin says that species evolve through the process of descent, these being modified giving rise to other species. "

Regardless of what science or the Holy Scriptures assert us, we still wonder what the real purpose of our stay in this existential plane is. If we were formed from the dust of the earth as described by the author of Genesis or if we come from the gradual evolution of microorganisms that arose from the cosmic explosion that occurred millennia of years in outer space, as well as if in fact we descend from certain species like the monkey still so there is a huge void that prevents us from understanding why we are here.

Why are we born? Where did we all come from? If we are spiritual beings surrounded by a physical body that is born, grows and then dies, what is the real purpose of this? We certainly don't go through this whole process for nothing, something must be important about it and we need to find out what it is.

Unlike scientific theories and biblical concepts that tend to clarify how we were created, only Philosophy is entirely dedicated to wanting to explain to us why we are here.

It is she who strives to explain our need to exist, our deepest values, thoughts, feelings and dreams. From this it is possible to consider that we are following this same step, the same purpose, wanting to understand so many why. Did we come to this world just to live our simple little lives and then return to the other side? Or have we all been placed on this ground where we stepped for a specific purpose?

What is the reason that only a few of us go through this existential plan and leave its own mark on it that makes them immortal due to their great deeds? Think of great scientists like Darwin and others of similar importance for the intellectual development of humanity, true geniuses who discovered or did brilliant things like electricity, technological evolution, those who changed the course of things.

Let us take into account people who have not been in our midst for hundreds of years, but their names have been recorded in history because they were capable of great deeds never forgotten by humanity until today and will continue in the centuries to come. Which of them stood out so much in their time, are immortalized and others simply passed through the land and did nothing interesting, as a result of which their names were erased even from the memory of those who knew them well?

In most cases someone is born, grows, dies in such a mediocre way that not even his family members remember him. Because of this, it is extremely important that each individual in particular endeavors to accomplish something that they are proud of, that piques the interest of other people for themselves so that after their journey to the afterlife their name is not buried together with their physical body. in the pit of oblivion. Many of those who care little about this significant detail and are now forgotten, no longer remembering their existence.

What is the use of coming to this vast planet if we do not do something for the evolution of ourselves and humanity? Religions call our attention to the fact that God wrote a story where there are the good and the bad, the light and the dark, the final judgment where each one will receive what they deserve, salvation or condemnation, the sects point to evolution of the spirit. Science studies ways and methods to evolve human knowledge in different areas of life in various ways.

Who we are? — About the Mysteries of Our Existence

Through research ranging from technological advancement to combating deadly viruses such as Covid-19. Philosophy, in turn, contributes a lot to our intellectual evolution, as it highlights the importance of each person in particular trying to discover its value in this life so that it can contribute significantly to the knowledge of ourselves.

This is vitally important because if we know who we really are and the real reason we are here, it will be easier for everyone to follow their own path and fully fulfill their mission. All those who once accomplished great deeds and today have their names on the walk of fame, became immortalized, unforgettable, marked their generation because they were able to know themselves deeply, they looked with optimism towards the future and bet with confidence in your ideals.

Only optimists reach the top, only those who trust God, themselves and their dreams can climb the highest mountain of their personal achievements. However, how can we do this if we are not aware of ourselves? If we are not sure what we came to this world for, what is the real purpose of being here? Let us always ask ourselves who we really should be so that we can find an answer that will lead us to finally discover the foundation of our existence.

I have analyzed the history of great names in the past, ranging from men like Moses, considered the greatest leader of the Israelite people even though he lived under such pressure; David, who ended up becoming the most important king of Israel despite apparently being unworthy of reaching such importance before his people.

Jesus Christ who marked the history of mankind by making a sacrifice for the liberation of the captives of Satan, reconciling man with his Creator. Great explorers of the Christian faith, ranging from Paul to those who gave their lives in favor of the spread of the Gospel.

And religious advancement from the first century until the Protestant Reformation that allowed the advancement of Christians even today. Men and women who, due to their incomparable intellectual abilities, allowed the world of today to be incredibly evolved.

Those who brought civilizations to know about electricity, radio, steam ships, the possibility of flying, of going into space and visiting the moon ... Why did they manage to do so much in an incredibly backward time compared to the century when we live and we are not able to do anything that can leave our name engraved on a sign on the corner of the street where we live, in a book, in a documentary on TV that shows the great deeds that changed the course of humanity?

It is certainly because we have not yet taken due account of how necessary it is to know our true potential, we still do not understand that we are not passing through the land just to wake up every day very early and go to work, meet with our family at nightfall, with friends over a beer round during a weekend barbecue, going to church to pray and confess our sins, dreaming of a car of the year or a better house ...

There is something more grand that we must do, something that makes us special, that will mark history and allow that even after death our name will never be forgotten. It was with this in mind that I decided to become a writer, nothing can make a person as simple as me than art, whatever it may be, and literature immortalizes us. Very early in life I carried within me the desire to keep my name in the history of mankind even without understanding how and why.

I wrote my first poems in a notebook and started to donate to the girls at school and won many hearts with it. However, my main goal was not this, what I really wanted was recognition, but it was only after the age of forty that I started to write professionally and today I have works sold worldwide in three languages.

Who we are? — About the Mysteries of Our Existence

Of course, I will not be hypocritical to the point of saying that I am not looking to evolve financially, because one does not live in this world without money, but at the same time I wonder what would be the use of conquering the whole world after dying and being forgotten? I do not want people to forget that one day I lived among them, I do not want to pass through the land without leaving my mark, to be just another one who occupied a small space among men and then ceased to exist without any importance.

During my walk I have met many people who have fought, conquered impossible dreams, enriched themselves, built large or small material empires, but in the end they died and are not even remembered by family members despite having left them an enviable financial structure. In my opinion, having a lot of material possessions in this life is not everything, I value what an individual will do to make his name stand out among the immortals of his time than his countless treasures.

When world-renowned artists such as: Elvis Presley, John Lennon, Michael Jackson, Whitney Houston, in addition to the Hollywood celebrities lose their lives the whole world cries and mourns their loss. This is because they have marked entire generations with their art of singing or acting and if we look carefully we will see that they came from the bottom to the top.

We can highlight Michael as a man who, even though he was born into an enormously miserable family to the point of having to collect food scraps from the city dumps where he was born next to his family, managed to stand out among the other brothers and became the greatest musical icon of all. the times.

Even after his death his album Thriller released in the eighties is still the best selling of all time and despite having become one of the richest artists in the globe his fame was what immortalized him.

Looking at Michael, I wonder if it is not true to say that we did not come to this world with an important mission, he brought in the package of his existence the task of entertaining millions of people with his art of dancing and singing. His name was recorded and immortalized in history because he gave himself body and soul in search of self-knowledge, he discovered himself as a great artist, singer, dancer and took the realization of this very seriously. Based on this reasoning, we can ask ourselves about what we came here to do.

Is the biblical concept that we were born just to grow and multiply our seed the sole purpose of the human race? To fight day by day for a better, healthier life, to accumulate assets, to gather lots of money, to raise sons and daughters, to form our families, to eat and drink while enjoying our friends, to acquire a good house and to buy the long-dreamed year ... Is all this enough? Is this just what we should look for in life? And when this life is over, will it remain of us? Will we be remembered or forgotten?

In the teachings of Christ we learn that we are not to seek glory and fame among our fellow men, but we can observe that even without this purpose he himself became immensely famous and his name has influenced millions of disciples for more than twenty centuries today. Reading what the prophet Isaiah wrote four hundred years before his coming, we see what his mission was.

"I, the Lord, have called you in righteousness, and will take you by the hand, and keep you, and give you as a covenant of the people, and for the light of the Gentiles. To open the eyes of the blind, to remove prisoners from prison, and from prison those who lie in darkness "Isaiah 42: 6,7 This was Jesus' main mission, he was sent to this existential plan to announce the Good News of salvation to man in order to make him recognize his sinful state and seek to be reconciled with his Creator and as we can see he fully exercised this commitment.

Who we are? — About the Mysteries of Our Existence

As we can see, the Messiah came to convince sinners to abandon their evil ways and turn to what is divine, he faithfully exercised his role and over the centuries many other characters whose names are written in history have led mankind to grow in all senses, both in the Christian faith and in the arts, science, technology and in various areas of knowledge. I am currently writing this book whose content will awaken my readers to understand themselves.

And you, what have you done or intend to do to impact those around you or reach the highest point of the human race? How do you plan to contribute to a radical change in history? Or is he content to not be noticed, carrying in his mind the mediocre thought that it doesn't matter to be remembered after he leaves for the other side? Perhaps you follow the Christian idea that we should live hidden behind the palm tree without showing those who are watching us how capable we are, but think carefully:

If people like Jesus and countless others had thoughts as small as this one, we would still live in the times of wax candles and riding horses. If today we have electricity to light our homes, power our appliances, we travel the land in luxurious cars and fly like birds, we modernize, we evolve ...

All of this was only possible because one day someone dared to think differently. And know that all this human wisdom does not come from man himself, but who gives us is the Creator of life himself at the moment when he decides to send us to this planet. Sacred Scripture itself warns us about this, it states that God blew into the clay doll's nostrils and placed a spark of his spirit in him, making him a living being, his image and likeness, endowed with knowledge.

It is logical that not everyone has the same gift, the same talent, but they are undoubtedly equally useful and important in the process of evolution. Be it in the arts, music, dance, science, knowledge, technology, politics ...

Each one in his own way, in his own way, according to his ability. Let us look within ourselves and ask: What gift did I receive at birth? What have I been endowed with since maternity? What is my importance in the change process? How can I be useful for the evolution of myself and others? We must keep in mind that we cannot live as irrational animals live only in the search for daily sustenance, a place to lay our heads and procreate.

But shake off that inertia that prevents us from growing and doing great things so that let us make our long journey something to be proud of in the future. When commenting on this topic with some friends I always mention the case of a man I met years ago who had the potential to become an excellent mechanical engineer. However, due to his constant inertia and neglect, he ended his days of life working as a simple metallurgist without any professional relevance. Skilled in the art of creating machines and designing electrical circuits.

He built magnificent equipment that not even those who for years studied the profession in colleges were able to do. However, he used to say that he did not study because he was not born to pursue a diploma, he preferred to dedicate himself to hard work than to walk around with the title of engineer.

Honestly, I am devastated to see so much talent being wasted, because the gift we bring from the cradle is exactly what we were born with the mission of exercising during our passage through this arid desert where we often find ourselves, there is no point in blaming God or destiny if nothing we are or we possess because if we don't act we are the only ones to blame. What kind of talent do you have, dear reader?

Does your voice have an excellent voice? So sing professionally, try to make a career, launch your CD, show everyone the gift you received from life. Can you create interesting stories that friends and family often love to hear?

Who we are? — About the Mysteries of Our Existence

Do you know how to fantasize? Become a successful writer, take it to the world and become a Best Seller like so many others have managed to be! Do you feel empowered to represent? Make theater, try a spot on a TV channel, start to shine! Conquer your space as an actor or actress, be part of the celebrities of your time, put your name on the list of fame. Is it good with a ball at your feet or in your hands?

Run after a place in a club, if you can't make it happen, show yourself, allow someone to recognize your talent and open the door to a great opportunity in the world of sports! What counts here is not to stand still, hiding within ourselves the gift and talent that we bring in our interior, as we do not receive it for nothing. Remember the parable Jesus told his listeners about the talents in the Gospel of Matthew chapter twenty-five verse fourteen, where he explains that three servants each received a certain amount of money to work with and multiply them.

However, only one of them did his duty while the others relaxed, so they were penalized with death. In the same way it will happen to us, because life endows us with gifts and we often do not endeavor to put them into practice, so we fail in all attempts to progress in other areas of our existence.

Because the path proposed by the destination for our victory has been rejected. . Do you remember the story I told earlier about the man who had the potential to become a splendid mechanical engineer and ended his days in complete poverty without any professional expression? Well, that's exactly what I'm talking about, If we don't try to win professionally through the gift received as a gift from the Universe, from Nature, from God or whoever gave it to us, we will face the wall. In fact, there seems to be a certain mystery about this, as it is perceived that the world has the need for certain hands of works and when we are born we are endowed with this capacity exactly to fill this space.

And, when we refuse to do this, we perish professionally. There are a lot of people who brought the gift of teaching inside and could have become an excellent educator, but considering the profession mediocre, without any expression, with very low earnings and little recognition on the part of those who pay their salaries, they chose to choose other areas However, academic training did not reach prominence, prestige or even a place in the job market, which forces them to exercise functions that are alien to what they were really looking for in life.

Contrary to what they did, others followed their intuitions, believed in themselves, started teaching in small study centers, then arrived at a university and finally founded their own schools and today they went from teachers to successful entrepreneurs in the field of education, this reflects the truth that we should not stray from our destinies.

This clearly reflects the idea that we will really be in life what we firmly believe to be capable of being, in this way, whoever does not aspire to be great will always be small. People who dream big climb the summit of the hill and are dazzled by achievements that for others seem impossible, but those who look tiptoe without courage or willingness to move forward confidently, willing to face all challenges and obstacles until they touch in the clouds it will never prosper.

It is not possible to accomplish great things in our lives if we do not wake up every day believing in our own potential, let go of the fear that what we want to have or do is something unattainable, beyond our reach or that for some reason we do not deserve to own and conquer.

If we were born on this planet of enormous possibilities, it is because whoever put us here wants us to do great things. Everything in this world is ours, just that we are willing to do our best to have them.

Who we are? — About the Mysteries of Our Existence

Optimistic people are known to be positive because they are never afraid of any obstacle that may try to prevent their achievements, they are strong, determined, self-reliant and never give up on what they seek to achieve in their lives. The negative ones, however, are the opposite of them because they have no faith in their own potential, are weak, pessimistic, fearful and live murmuring because of their constant defeats, they look to the future without any form of hope.

There is a popular adage that says that life is quite hard for those who are soft, that is, whoever kneels in the face of the struggles that the world proposes to us will be overcome by them and will be a constant failure. It is very common to see individuals who claim never to have had great opportunities and for that reason it has become impossible to achieve the fulfillment of their dreams.

However, we know that in the end they gave up chasing their goals after the first fall. Make no mistake, existing in this world is not easy. If at some point we have already asked ourselves who we are and why we are here, the answer is that we are privileged people, because we are given the merit of God or the Universe to be chosen to live on this earth in order to become explorers, conquerors, filmmakers. great achievements.

On one occasion I heard an evangelical pastor say that the angels wished they had received permission from the Creator to go down to where man is lost in his crimes and sins to announce the Gospel to him, leading him back to the path of salvation, but this mission the church was given and not to them.

Look how interesting, according to the words of that evangelical leader, God gave man himself after the liberation of the right to evangelize those who are still captive to Satan, despite being imperfect. If we analyze carefully we can see how special we are.

First because according to the Bible we were made in the image and likeness of the Most High, then because we were placed to live on the most important planet in the galaxy because it is the only one that allows human life with all the resources that we need. In front of us there were several opportunities for conquests in the form of dreams, aspirations, challenges, goals to be overcome and these things are what gives life meaning. We need to reach the peak of the mountain!

Mount Everest is considered the highest mountain on earth, its peak is 8 848.86 meters above sea level, a huge number of people have already challenged their own limits when trying to reach the top of this mountain and prove themselves who were able to overcome any obstacle. In the same way, we must all act in relation to everything in life.

The fear that we sometimes have to fail in the face of supposed obstacles that arise along the way is responsible for most of our setbacks, because in the end they end up preventing more positive results in what we try to accomplish. Whenever we set a goal for the future, we need to be aware that everything is possible to be achieved. Jesus Christ once told his disciples that everything is possible for those who believe. Well, humanity's greatest spiritual leader was not in the habit of lying.

If there are one or more neighbors on the side of our houses who have already obtained their dream home, a luxurious apartment, their brand new car, it is because they certainly ran after the fulfillment of these dreams with faith and optimism, for this reason they were successful.

And if by chance they got there, we can also do the same if we are determined to try as hard as our neighbors did, fighting, paying the required price, giving up that, working harder.

Who we are? — About the Mysteries of Our Existence

Saving, leaving aside amusement and excessive spending, if we refrain from leading an exaggerated life and bet all our chips on the realization of a greater good, we will certainly succeed in what we set out to achieve. No one is born to be a loser, if we are no longer or we are not, it is because we spend most of the time admiring or feeling envious of those who have already reached the top of the mountain without first trying to climb the hill together with those on the way.

It is now time to change our concepts and move towards our dreams with more boldness, faith, confidence, hope and without the terrible fear that only serves to lock our feet with the chains of failure. Let us get up, let go of the old fear that told us it was impossible, raise our heads and with a more optimistic look let us say to ourselves that we are capable.

The history of mankind was not and will not be made by fearful and weak people, because they do not leave their mark on human existence, they pass through this world completely unnoticed, without the slightest projection or importance.

So that our passage through the planet earth is not a simple shadow without the least expression, so that our steps can be marked in the sand, our name is not forgotten and we do not arrive on the other side empty-handed before the one who sent us, greater determination is needed .

There is no longer any place for those who think that their existence here was by chance, it is nothing more than an error of nature, which resembles the irrational animals that are born, grow, die and become dung. The man is formed by body and a spirit that one day will leave matter and cross from that reality to another and there he will surely give an account of all his acts practiced on this side of life to those who proposed the possibility of him to exist in this world. We are beings sent to make the history of mankind happen.

Some invented spectacular things that changed generations, others made incredible discoveries that revolutionized the way of acting and thinking of all civilizations on earth and now it's up to us to make our generation step forward, evolve, grow, modernize more and more . It doesn't matter where or how we do it. It can be a social work that allows your community to prosper, help the most needy, provide humanitarian aid, act in health, in politics, in social actions ... Let us do anything that changes the course of history.

Whether we are part of big or even small changes that allow us to contribute to a better future, let selfishness look only at our own navels and start doing or creating something in group, seeking a greater good, the collective prosperity of a street, a neighborhood, a city or the whole world.

If it is correct to say that I will always be what I think I am, then I need to think with more responsibility so that my life is graceful, exuberant, fruitful and full of good results in everything I do. Enlightened people bring lights to those who observe him, progress will be his mark wherever he goes, his steps will leave deep marks where his feet will stand, they will be the reference of dozens, hundreds or even thousands of others who will follow him, but he cannot be all without a brilliant mind.

Whoever spends all the time with their thoughts based on a constant positivism will progress towards a future of excellent results. But, on the contrary, whoever has a limited view of life will not move. What I am trying to make clear from this the first words written in this book is that there are two types of people, those whose lives bring about significant changes both for themselves and for the evolution of others, whether in a broad or limited sense like the family, a small community ... And those individuals whose existence is so insignificant that they have no impact on global, local or self-evolution.

Who we are? — About the Mysteries of Our Existence

Once again, quoting from the Holy Scriptures, I want to quote here the story of Peter, a disciple very close to Jesus who swore fidelity to him above anyone else there, however he denied it when he was pressured by the Jews. The anguish of that man when he saw his Master fall into the hands of his enemies and perish on a rough cross caused him an extreme anguish.

Peter really wanted to be faithful to his friend, that desire really existed in his heart, in his heart, but he was not prepared to face the pressures that came after him. It was necessary that some time later he came to believe that he was able to fulfill this desire with more tenacity and confidence regardless of the price to pay.

We read in chapter three of the book of Acts of the apostles the wonder that that apostle became after his sudden change of attitude, his thoughts changed and then came great results in his performance as a disciple of Jesus. When he went up to the temple he performed healings, deliveries, his words were more convincing, more firm and powerful, in this way he convinced about three thousand people to accept Christ as Lord and Savior of their lives.

Like Peter, we will all see great things happen in our lives and in those who watch and listen to us when we start to think more clearly, believing in God and in ourselves, in our ability to conquer, to achieve, to realize our ideals. . Each one's inner world can be enlightened, prosperous, evolved or remain in spiritual darkness, miserable, fruitless, arid as the densest desert, everything will depend on how we conduct our actions and thoughts.

Who are we today? What are our deepest thoughts? Are we visualizing good or bad things in our minds? Are we believing better days or does pessimism only lead us to believe that the future will be even worse than the present? Are we optimistic or pessimistic people about what will happen to us tomorrow? Do we still have hope or have we lost faith in God and our fellow men?

02 – WE ARE RESPONSIBLE

1. RESPONSIBLE FOR WORLD PEACE

Since the beginning of the centuries, civilizations have been divided between those who seek world peace and those who profit from wars. From Jesus Christ to current religious leaders the message that man must abandon violence and live in the most complete harmony with his fellow men has been preached, however, history shows us that only the first century church really sought it sincerely and paid a high price for that where some were beheaded, burned alive, hanged, nailed to crosses, delivered to the beasts.

In the centuries that follow to the present day, religions are nothing more than stagings where their leaders use artifice to deceive the least understood and take everything they can to enrich themselves easily. There are many entities around the world that call themselves defenders of animals, of the poorest, of nature ... However, we know that everything is just a band to hide the real pretension of these false peacemakers, because the real goal is to get along and accumulate great wealth.

Man considers himself modern, evolved, but continues with the same destructive characteristics of the principle, his tendency to evil, violence, to fights, the pleasure in the destruction of his own species remains unchanged in his DNA. Human wickedness has manifested itself since its creation and continues to exist within your heart to this day. What is the use of talking about peace while building increasingly lethal weapons in order to destroy your fellowmen? If, in fact, man wanted to find the peace that he seems to want so much.

It would be enough for everyone to let their guard down and stop destroying themselves, abandon this constant search for power, the ambitious will to want to be superior, to dominate other peoples by force, to want to step on over the head of the weakest. Peace is a state of mind that requires unity, understanding, dialogue, absence of hypocrisy, love and forgiveness. Worst of all, it is exactly these qualities that are lacking in human personality.

I live in the region with the highest rate of agrarian violence in the country, here the dispute for land goes beyond any humanitarian limit and hundreds of people have already lost their lives because of the ambition of people who, for not being content with what they have, take possession of other people's properties. at any cost, based on violence, violating all laws and rights without being held accountable by the Brazilian justice, which is slow and corrupt. In this country where money speaks louder, the powerful do the damage they want without fear.

And before that still appear those who to convince the nation that they are good, desiring the support for their candidacies for political positions, who dare to appear in public declaring themselves as defenders of global pacifism. Bunch of hypocrites, that's what they are. We are all responsible for violence, for the misery of others, for the parade of death through our streets, for the amount of blood spilled, for the lack of hope, for the growth of corruption, for the advance of wars and struggles.

We are to blame whenever we support those who commit these things and remain silent as a sign of neglect. When we act violently against someone, when we do not know how to forgive, we refuse to forget an affront, we refuse to extend our hands to the hungry, we lie, we betray, we hurt our fellow men. If we want to contribute to peace, we must practice love, be more tolerant.

Who we are? — About the Mysteries of Our Existence

Violence begins within ourselves and externalizes from our actions, hence verbal, physical aggressions, extreme acts against anyone who will also react in the same way, leading this explosion of hatred, resentment and dissatisfaction to dozens , hundreds and finally thousands of people as a deadly virus that spreads across the planet, causing the death and degradation of mankind. The anger that burns within our hearts can destroy the world we live in like a spark of fire that spreads.

It falls through the dryness of an arid land and is capable of burning large forests, putting an end to the lives of countless animal species and transforming cities into dust and dust. Because of this, we also become responsible for the wars and struggles that occur in the four corners of the planet in which we live, when we are not able to collaborate with the world peace that we need so much. If it is true that the Universe responds to each of our thoughts and attitudes, then we must take responsibility for acting and thinking.

It is because of our bad actions, bad procedures and sayings, the irresponsible way we make certain decisions in life, the injustices practiced, the acts of hypocrisy, the lies, the mistakes, the corruptions and ambitions, the creation of laws that only they harm society, which today we look around and what we see is the increase in violence on our streets, poverty and misery consuming our people, social inequality as a whole, the immense lack of hope that leaves us unsure about tomorrow.

Unfortunately, we ourselves are the cause of so much misfortune that it ends up increasing in size every day to the point of covering up and suffocating us, like an avalanche of great proportions, a strong storm that comes at an uncontrollable speed, taking everything it finds through the front. Our actions, individual or collective, determine good or bad results.

2. RESPONSIBLE FOR A MORE FAIR FUTURE

What will our tomorrow be like? Will we have more peace and tranquility than now in the present where we find ourselves? What will be the future of our children, of the next generation, of those who, because of us, came to exist in this world? Let us look at least for a second at the children who play and have fun inside our houses, in the squares, woods or somewhere destined for leisure. Let us see the tranquility in which they sleep, their innocent smiles upon waking.

They are calm, confident that we as adults are working to guarantee a more peaceful, safer, more peaceful tomorrow. But, as your parents, are we even having this responsibility to guarantee you a better future? Does simple care give them a good education, a happy and comfortable home is enough so that ahead of them they can live on a less violent, polluted planet, consumed by social inequalities, corrupt and hopeless?

If each one in particular plays a part in cooperating in the existence of true peace, planting true love in their hearts by treating their fellow men with more respect, dignity, seriousness, honesty before their eyes, they will acquire an identical character that will lead them to act in the same way after reaching adulthood and as a result of this long-term investment the world they will live in later will be formed by the most lovable creatures of themselves and this will be reflected in an unparalleled peace.

We all want love to spread across the land so that people stop destroying themselves, that death stops staggering through the streets of our cities, the corrupt disappear, the poorest are finally remembered and receive the help they deserve , fairer laws are created, inequalities cease to exist ... But we do nothing to make that happen.Tomorrow may be better than today, but it is necessary that we are willing to hold hands and in one purpose, walk in the same direction.

Who we are? — About the Mysteries of Our Existence

T5owards the definitive practice of sincere love for each of our fellow men, practicing justice with those that today's society has discarded, thrown into the gutter, showing mercy and pity to those who suffer pain and neglect. Only then will the Universe understand how much we need such a great transformation and change our destinies. We democratically chose our political representatives on the pretension that they can create laws that benefit society as a whole, however, what they do is give more to those who already have too much.

And take away the little of the less favored, those who have hands the power to change the course of things in this world care little to act honestly, contributing to a better future for all of us. So, if these people refuse to do what they have been called upon to do, you and I, ordinary citizens. Renowned scientists in the scientific field say that our fate is uncertain and that in a few centuries the human race will inevitably be decimated.

Space research centers like NASA are already searching vast parts of outer space in search of a planet that presents the same conditions of life so that it can be inhabited by those who can afford to pay millions for a place in the new land and thus guarantee their livelihood and that of their family there, when the end finally comes to us. But, is that perspective really correct? Will our planet be decimated in the future, or is that just scientific superstition? What matters at this moment is not to believe or not in these concepts of the end of the world, but to start planting the seed of love, charity, mercy and to sow justice among ourselves so that we can guarantee a future peace.

3. MORALLY RESPONSIBLE

The world we live in is morally corrupted in all its social areas, whether political, religious or family. It is the duty of each of us to cooperate in the best possible way so that this sad situation…

Can be reversed or at least minimized its impact on our lives. It turns out that most people do just the opposite, contributing to further increase the advance of global immorality. Homosexuality has become something accepted as if it were a natural condition of the human being, but deep down we know that we are not born to invert our sexual options and stop being what we really are, women are born to be sexually attracted to men and, these, For them.

The truth is that most individuals who support this shameful condition of humanity refuse to understand that a gay or lesbian is not changing sex, even if some change their members, deep down they just lose their true identities.

Once, staring at a transvestite, I wondered who my eyes were fixed on, whether it was a man, a woman or anything without the slightest definition among the human race. Because that is exactly what happens to someone who tries to physically transform into another person, he just disappears and gives way to something strange, without logical identification. This is what homosexuality causes in people who adhere to such behavior.

On the other hand, sexual immorality also intensely destroys the human character, leading individuals to destroy themselves and those around them, especially their own families. Men and women who commit acts of infidelity put an end to their marriages, hurt each other, sadden their children, ruin their homes. Those who choose to sell their bodies in a motel bed and are considered girls or prostitutes are at risk of serious illnesses, in addition to losing their values as good people.

They start living on the margins of society and even if they earn a good salary for this practice they are part of the social scum. Prostitution is harmful to human beings both physically, morally, socially and spiritually because it distances us from God.

Who we are? — About the Mysteries of Our Existence

It is not possible that a nation can grow morally if its inhabitants prefer to live in the likeness of the former residents of the cities of Sodom and Gomorrah, quoted in the Scriptures.

Where immorality was so great as to be destroyed by fire and sulfur fallen from the sky. It may be that for the vast majority this event was nothing more than a legend, but if we look closely at the present we can see that it is due to so much moral rot that the modern world has a bad smell. The sin practiced by human beings in this world manifests its bad odor through violence, hunger and misery, injustices, corruption, so many epidemics, viruses that kill millions of people ...

In a way, nations are being destroyed, paying a high price for his boldness in continuing to defy the laws of the Universe. We are, yes, morally responsible for refusing to satisfy the filthy desires imposed by our bodies, avoiding satisfying it through acts that only embarrass and demoralize ourselves. A more just society will certainly be more blessed, more prosperous, less violent and its evolution will be noticeable.

Otherwise, if you insist on a promiscuous life you can even grow economically, reach high technological levels, become a great world power, be among the richest countries in the globe, however, there will be no peace among its inhabitants, its enemies will surround it, death will continually walk the streets of your cities and chaos will be permanent. Looking at the modern world, we can see so many nations that have evolved to the point that today they are considered the economic columns of the planet, however, from them arise wars.

Financial crises that affect the rest of the world, pests in the form of viruses that transmit to residents incurable and deadly diseases. What good is so much power if they end up being the root of evil for others and for themselves?

And why is that? For the simple reason that they abandoned their moral values, supported immorality, became fond of evil and turned their backs on good. Due to so many moral extravagances they attract serious consequences for themselves. Because the power of darkness is something real and usually acts with greater intensity where there is a supreme index of disobedience. Man is a being formed by body and spirit, this is a universal truth confirmed and accepted by those who present a minimum of intelligence.

In the 21st century, only the most skeptical deny the existence of two worlds, one material and the other spiritual. One formed by the darkness of sin and the other by divine light, it is up to each individual in particular to change his sinful state to adhere to his place in the clarity that resides in the beyond where he will go after he leaves this existential plane.

However, the two worlds intertwine with each other through mortal man, being influenced by the action of their acts practiced while living on earth. The more we sin, we commit evil, we demoralize our bodies, we yield to the evil that is already planted in our DNA, we allow darkness to invade the light on the other side. We open the way for the forces of evil to dominate our inner being, our spirits and envelop the entire planet we inhabit in dense moral darkness. Therefore, we must accept the fact that we are morally responsible and correct our mistakes.

4. SPIRITUALLY RESPONSIBLE

As I mentioned earlier, we have two existential plans, a primary one where we were born, grew and existed until the end of the journey. The other is secondary and comes only when we encounter physical death. And the doors of eternity are opened. The first is material and the second spiritual, however, are directly linked and are influenced by what we do or think about in the reality…

Who we are? — About the Mysteries of Our Existence

In which we find ourselves now. The material and the spiritual world coexist and are impacted simultaneously. Here it is possible to explain and understand why there are so many social ills, deaths, epidemics, poverty and misery in different parts of the earth, the revolt of nature in the form of earthquakes, hurricanes, floods, storms, global warming.

None of this happens by chance nor is it a natural process, everything is the result of man's constant action in defying the laws of the Universe — or of God — refusing to follow his directives permanently. When Adam sinned in Eden against his Creator, he said: Cursed is the earth because of you!

All of our actions and attitudes practiced on earth will have a negative influence on it, since it is sin that destroys it. According to what we read in Genesis, we had been created to be eternal, immortal. It was the disobedience of the first couple that changed this condition of human beings and death came to limit our life as a form of punishment for our transgressions. Let us observe that even the planet itself was condemned together to suffer the influences of sin through our actions.

As we are an integral part of two worlds, where one reflects our human form and the other our spiritual essence - who we really are - we end up being responsible for what happens on both sides. If we are lovers of sin, our two worlds will be dark.

It is common to hear someone say that a certain person is negative, pessimistic, who never advances in life. These individuals are at a late stage on the spiritual plane and for this reason live with their physical feet and cannot evolve. For the physical world to enter the process of evolution, it is necessary for humanity to strive to improve its spiritual side, what we commonly call "our inner self". This Inner Self is nothing more than our true essence.

What religions call the soul or spirit. In creating man, God breathed into his nostrils a spirit of life that made him live. That divine breath is us, according to the biblical account. But, what does science say about our origin? Well, as I quoted at the beginning, if anyone wants to consider the idea of the monkey or any other micro organism feel free, I prefer to believe that we were created by the hands of the great architect of the Universe.

How to prove that we are beings formed by body and spirit? Let us stop to think about how we lay in our beds, sleep and in the blink of an eye we are elsewhere through our dreams. How can we explain this ability that we have to leave our physical bodies momentarily and go for a walk in other places, talk to other people, live new experiences and even go back in time to relive old memories?

Well, science tries to clarify this phenomenon by giving the idea that dreams are just reflections of what we live in day to day, think or desire.

In light of that, I myself had an experience recently, spent days reminiscing about things from my past. I fixed my thoughts on things that I wanted to happen in the present just to see if I would dream of these things, however my dreams — which usually happen frequently — never pointed in this direction which led me to believe that scientific theory in this regard is a mistake.

Spiritist doctrine, on the other hand, says that dreams are the exit of the spirit from your physical body for a short period of time to go out to venture elsewhere and keep in touch with other spiritual beings while your physical form rests from daily work. . And seriously, I really believe that because I believe in my spiritual essence. I don't see myself just as a being formed only by a lot of flesh, bones and blood, I feel that I am something beyond matter. I do not agree with the limitation of the human being to a mediocre and finite existence.

Who we are? — About the Mysteries of Our Existence

There are things that the human mind cannot assimilate or explain clearly, but that does not mean that they are just the fruit of our imagination. In my way of thinking, I understand that if we came to this world it was for a special reason, perhaps to create something of great significance for the human race, to make the world, the planet and even the Universe evolve more and more. So, if so, why would we be finite?

No, death takes away our physical bodies, but our essence continues forever. Hence the certainty that our importance in the evolution of our species is vital on both the material and spiritual levels, as we are part of both sides. People who believe in this endeavor not to go through the land without doing or creating something of extreme value in their lives, they constantly seek to stand out among the others, they seek to shine among others who remain hidden in anonymity.

They strive to leave their names registered in history before leaving for the unknown. I have in mind that the theory of reincarnation is something true, this central point of the spiritist faith convinces me that our life time expires and we leave this world, going to the afterlife, however, afterwards we return occupying a new physical form, being born in a new womb, in another part of the earth, in another social, physical, financial condition, to continue what we started.

If this reasoning is really true, we can understand that human evolution depends on the spiritual growth of each one. Anyone who goes through this existential plan and does nothing interesting to contribute to its expansion in any of the areas that life offers him will die one day without taking a single step forward and his coming into this world was in vain. We are not just here to live our petty little lives, but to push our species towards great achievements. Many who have preceded us have already done this, let us look around us and see so many new things, incredible things created or discovered by brilliant minds.

True geniuses! Could it be that only they were given the spark of divine and universal wisdom to be able to fulfill their evolutionary missions? Would the Creator of human life or the Universe have become selective to the point of condemning most of its creatures to a mediocre life, while only a minority would be given the gift of helping in the evolution of humanity? Of course not! We are all born with a specific mission, a gift, a form of wisdom, a talent, however simple it may be.

From the snake charmer in India to the great scientists who enchant us with their surprising discoveries, all are important in the process of evolution of the species and the planet where we live. But, for this to happen, we must first get to know each other better, grow in knowledge and wisdom, learn to unite our physical and spiritual worlds, grow in both existences.

Only then will we be able to evolve with more intensity and without the destruction that we see today. No more wars, deaths, corruption and all kinds of devastation that unfortunately we have witnessed outside the world. If people finally understand what they are and what they came to do on this planet, things will improve, we will find the peace that we have been looking for so much.

5. SOCIALLY RESPONSIBLE

There is no use trying to disguise and pretend that the social problems that surround us are not our problem, as they are real and affect all people directly or indirectly. The injustices inflicted on the poorest, the homeless, those who are on the margins of society and are treated with contempt by those who have the resources and power to help them. The corrupt who remain in power stealing our taxes without returning to us in the form of basic sanitation, health, public works, education.Everything that happens around us ends up influencing our way of living.

Who we are? — About the Mysteries of Our Existence

And only the most insensitive will be able to close their eyes and go on their way without worrying about seeing so much misfortune falling on other people. Unfortunately, this is a sad reality that is spreading around the world without the powerful bothering to at least try to remedy it. Many social projects created by politicians and philanthropic entities appear each year to change this situation, but they do not solve anything because they are nothing more than a facade.

These programs hide behind the real goals of the power-loving bosses that are their own interests to keep their high positions by deceiving their voters with the false image of good guys, they wear a mask of concerned with the population while in the background they just want to get along . We are responsible for what happens in our society as a whole because we are part of it and the consequences of these distortions also reach us at extreme levels.

Even the mountains of garbage on our streets harm us in a group because they attract diseases that do not save the face of those who are rich or poor, white or black, those who live in shacks or live in their beautiful mansions, in luxurious neighborhoods or on the outskirts. . It is the duty of all Brazilian and world society to ensure public health and not just our government.

Sometimes I notice that the population blames public neglect exclusively on the mayors of their cities, governments and the President of the Republic for doing nothing for the country, but they forget to look at themselves and ask themselves what they do to improve it. it.

We rarely stop to think about what we could have done so that our city or nation had certain social changes that would bring us prosperity. If each one of us planted a tree in a short time we would recover all the Amazonian flora destroyed by deforestation, in the same way if we do not spread the garbage on our streets we will have a cleaner city, without the criminal fires that farmers.

And even simple rural residents do every year we will unburden the environment with so many gases that burst with the ozone layer even though large industries still exist, expelling a large number of smoke. The use of bicycles instead of motorbikes, of the most modern and electric cars instead of those powered by gasoline or diesel would greatly reduce the harmful effects on our atmosphere that so much pollution calls for help.

We must stop blaming the big industrialists for the large quantities of smoke released into space, our government officials for the dirt of our streets, because there are people in the most complete abandonment, who live on the streets, sleep in the gutters, under the viaducts. All of us must contribute in some way to changing this terrible situation in our society, starting by not spilling rubbish on the avenues, avoiding unnecessary fires.

Reaching out to those who most need help, participating in social services that aim at a collective good, distributing food to the hungry. It is much more valuable to worry about our fellowmen and try to remedy the poverty that makes them go through so much deprivation than to waste time pointing our fingers at you, accusing A or B for this or that situation.

It is very easy to blame others for social disgrace, when in fact we are all equally responsible for human evil. We are responsible when we close our doors in the face of those who beg in front of our homes, for failing to answer the cry of a hungry child. When we can and do nothing for the sake of those who mourn their losses, their pains, their needs. Let us stop for at least a second and ask ourselves what we have done so far for the existence of a better world where people can live in harmony, possessing the same things, with the same quality of health, education and that every day can wake up and have breakfast, lunch and dinner at the end of the afternoon for themselves and their families.

Who we are? — About the Mysteries of Our Existence

No matter how simple it may be. Maybe you and I have this financial condition, we can get together with our families around the table and thank God for never having spent a single day without the food we need, but, who knows, well on the other side of our house there are others who have not even eaten yet. There was no morning coffee, they didn't have lunch, they don't go to dinner later, nor is there any expectation of tomorrow being different. But since most of us human beings are selfish, we care little about the hunger that burns in them.

Most people are content only to see their barns always full, their pots always full, their good clothes and shoes, their bank accounts overflowing with money, their prosperity day and night sitting at their doorsteps. Some even say that they are not responsible for the ills of the world and that they are not in need because they are blessed by God.

All right, I even agree with the idea that God may even have decided to bless some and curse others, after all, through the mouth of the prophet Isaiah he himself said that he would have mercy on anyone who wanted to have mercy. He is the Almighty, the owner of the Universe and of our lives, so he can do whatever he wants. But in the book of Leviticus, written by Moses, he says that the Israelites were never to abandon their needy brothers, as there would always be poor people on earth.

Based on this principle, we can understand that even though he placed the rich and the poor on the same land, his intention is that the most financially wealthy reach out to the most needy and not that they despise, humiliate, knock with their doors on those who face them. ask for help.

It is not that we do not have social responsibility with the poverty of others or that we have all the best because God loves us more than others, we simply receive too much to help them in their needs.

WHO WE ARE? We are socially responsible to fight against this corrupt political system that allows only a small percentage of our population to have access to a quality life, their children study in good schools, eat well and live healthier. It is our responsibility, both mine and yours, to try to change this sad reality that has led this world to be so unequal in the way it treats the vast majority of its inhabitants, to fight for greater equality on this planet.

6. HUMANLY RESPONSIBLE

Wagner Dias Ferreira, lawyer and member of the OAB in the State of Minas Gerais, thus characterizes the responsibility that each individual has with his own humanity and with that of others around him:

"Responsibility is a value in society that needs to be cultivated in order to grow and produce human values that are important for people to live together. In law, when talking about responsibility. One can immediately think of illicit conduct practiced by the President of the Republic, the so-called crime of responsibility. Or, in the most common, civil liability, more linked to the daily lives of Brazilians, the people who with their feet on firm ground support the nation. Now, the responsibility of the supplier of products or services and often in strict liability, that of the State for the acts practiced by its agents or concessionaires. So what is responsibility?

At first, it is possible to observe how the word has the same radical as the word "answer". It can therefore be said that responsibility is the ability or ability to respond to certain situations. Law 1.079 / 50 defines several conducts contrary to constitutional commands as crimes of responsibility, demonstrating that, if the president does not respond well to compliance with the Constitution, he must be removed. Civil liability calls upon all people to be accountable for their own actions.

Who we are? — About the Mysteries of Our Existence

The responsibility of the supplier of products or services calls him to answer for the products he has manufactured or the services he has provided. And strict liability imposes on the State to answer for the errors of its agents or concessionaires, even if afterwards there is a return action against them, the State responds. So, responsibility is a value that needs to be cultivated and needs to grow and take on more space in contemporary society.

Everyone is called upon to respond when facts that undermine Brazilian society occur. This is critical to the nation's credibility. Your responsiveness is something that should be noticed. A people needs to have responsibility, and this ability to answer the questions that life poses must be passed on to generations so that the future is not chaotic.

Prophets were not people who divined the future and prophesied their visions or divinations. Prophets were people with an enormous capacity to answer current questions and preserve this ability to answer for the future. When they discovered the Dead Sea papyri, there was great unrest in Western religious communities. There, they found records of the Essenes demonstrating that the transmission of the biblical texts, mainly of Isaiah, had been faithfully carried out for more than a thousand years.

Isaiah was a prophet who described hundreds of years before many events in the life of Jesus of Christians. The most important thing with the discovery of the dead sea papyri, therefore, was not the confirmation of the prophet's texts, nor the content of his prophecies that were already known.

It was the demonstration of the responsibility of a community with the preservation of its knowledge, and transmitting their ability to respond to situations they faced responsibly. In recent times, many challenges of an environmental nature.

Produced by humans and produced by nature and, recently, public health like Covid-19, are calling on everyone, worldwide, to give answers, individual and collective, that promote the ideas of preserving and transmitting the memory of these responses to future generations. "

According to a report by the World Health Organization published by the BBC.COM website on March 31, 2021, the origin of Covid-19 is not yet known, however, it is known that: It is "probable or very probable" that there was an intermediate animal between an infected animal and man This scenario suggests that a first animal that developed the virus infected an animal of another species and the latter infected a human. This is based on the fact that the viruses found in bats related to SARS-CoV-2 have differences that suggest that there may be a "missing link".

Here we see the immense responsibility that we have with humanity in everything we do, whether it be a simple choice or decision in life as something of extreme importance as a scientific research. Due to the lack of care in the treatment of a possible virus in a laboratory research, it is possible that he escaped and infected some animal or even one of the scientists present during the research and later spread around the world.

Millions of people have died since then and so far nothing has been able to stop them from advancing across the continents, getting stronger and stronger with each passing day, killing equally from the youngest to the oldest.

When the Chinese government was warned of the incident in the laboratory, it paid little or no importance to the warning given by scientists, and it was only after the chaos had formed in the country that it realized the damage done. Basically, it seems that the government of that nation has found a means of all this catastrophe, a means of profiting from the global misfortune.

Since since then it has been earning millions of dollars in the sales of masks and respirators all over the world, it suggests that everything was armed in this regard, as we know that China has not won the position of the second largest economy in the globe for nothing, they usually act in a suspicious way for national enrichment.

Brazil itself has invested high sums of money there. In reality, all the great powers of the world find in social disgrace a way to withdraw their enormous economic wealth.

Americans, for example, amassed their greatness through World War II by being one of the last countries to confront those who were committed to Adolf Hitler's Nazis, so after the Allies had won the battle the United States was well structured economically and started to lend billions of dollars to other countries.

That was how they prospered, investing huge sums in heavy interest for other war-torn nations, earning unparalleled profits in rebuilding them. In addition, after this episode, he continued to invest in wars against anyone who would rise up against the greatest power on the globe or by financing the less armed nations to fight their enemies as in the case of Israel. World and social woes bring economic benefits to many governments and that is why nothing changes in this regard.

07. SEXUALLY RESPONSIBLE

Another factor in which we must be responsible as intellectually aware people is with our sex life, because it has serious consequences that lead humanity to achieve high levels of damage in various sectors of its existence.

One of them is the uncontrolled growth of child births in our cities, leading families with an extensive number of children to financial scarcity.

Where it becomes impossible to give them quality food, education and health, further multiplying social misery. Because of this irresponsibility, many young people and teenagers become unprepared to assume their roles as parents and pour on huge shoulders of family members, also without resources.

Huge amounts of children who are raised as time demands, without study or any other means that they can offer in the future. future the ability to become good citizens, useful for the growth and development of the nation where they live. And this poverty of knowledge leads them to live anyway, following marginality.

Marginal laws, created by politicians and authorities unprepared to govern a society lost in its concepts of existence, are largely responsible for the extreme disorganization and social perversion to which we all find ourselves victimized today. The Statute of Children and Adolescents (ECA), sanctioned on July 13, 1990, which was created with the objective of protecting them from domestic violence, ended up resulting in the debauchery of this class of individuals and diminished parental authority.

Today, aware that they are protected by this infamous law, boys and girls gather on street corners and steal, kill innocent people, parents of families, workers to steal their belongings acquired at high costs. After these insane acts, the reprimand of justice is to give advice.

If these child criminals, who afflict a society that is completely unarmed, abandoned and without the support of the competent authorities, were properly punished and condemned for their mistakes, surely the urban violence in our country would greatly diminish. However, while good people like you and I remain victims of child debauchery formulated by incompetent politicians, they remain in their offices creating new ways to destroy us socially.

Who we are? — About the Mysteries of Our Existence

Their families are protected in mansions guarded by strong security apparatus and the people are going to explode! They care little about our needs, whether we are doing well or at worst, for this group of corrupt people what matters is only the millions in their bank accounts and their financial prosperity at the expense of our taxes. However, it makes us wonder if we are doing the right thing, if we are also not responsible for all this social degradation. Let us analyze whether we are in any way included in the list of those who corrupt the system.

How many children do we have today? Were we prepared to assume our roles as parents? Are we correctly assuming this and other social responsibilities? Is the family we raise receiving everything they need to live well? Do our children go to a good school? Do they feed or dress properly? Or did we give life to several individuals and today, for extensive financial need, they live as delinquents, being part of the statistics of violence and death that fill the news?

Only after having answered these questions to yourself will any of us be able to condemn those who have done so in the past or at this time, because if we have the least share of the blame in these cases we have been irresponsible and deserving of punishment. The body responsible for informing the Neonatal data of the State of Pará, where I took up residence, reports that around two hundred children are born every day in Pará.

An alarming number that should awaken the authorities for birth control in this region and in the rest of the country, as a recent poverty survey was just as frightening. As we walk through the streets of our cities, we can see the immensity of children thrown around street corners, asking for alms at traffic lights, abandoned girls and boys. Therefore, I would like to know the opinion of those who defend the Statute of Minors and Adolescents with their thesis that our children should be free to choose their destinies.

As if they had the maturity to do so, preventing parents from educating them as was done in the past. , with short reins and discipline, what do they think of all this social disgrace that affects at least eighty percent of the Brazilian population, if not more than that. It would be ideal for laws created by a small group of clueless authorities to be overturned.

Because they prevent parents from raising their children within a discipline that really helps them understand and assume their true responsibilities as individuals who are part of a constantly evolving social environment. If we, as parents, can no longer rebuke or correct our children within tolerable standards, who will? The Justice? A social worker with advice that we couldn't give while they were growing up in our homes and that now won't have the least effect on them?

Scientific studies show that a child starts to have a complete understanding of things from the age of seven, from then on he already understands what is in fact right and wrong if we, as tutors, teach them properly. However, if we don't, they will grow up with their minds empty of wisdom and understanding and will only learn in adolescence.

And this is the worst period of human life as it is a transition phase between childhood and maturity, in this existential space the adolescent's mind is full of doubts, concerns that need to be correctly clarified and filled with guidelines that lead him to reflect on your rights and duties towards the people around you, the society you live in and the family.

If in childhood he does not have the basis for this, he will seek this information out of the house and it will happen that the one who will instruct him is the rotten side of the world. Most criminals today are minor offenders who, because they did not receive the necessary instructions in their homes.

Grew up in an inadequate way, without the slightest respect for their own lives, of the other people around them and did not fear the punishments imposed by a law that never allowed him to learn and grow properly guided by his parents. Here comes the worst of all, as the traffickers appear who entice them into drugs, then use them as mules for the transportation and sale of narcotics.

The first thefts arise, the comings and goings in detention centers for juvenile offenders where they hear the same bullshit of psychologists who think they can shape the mind of a young man who has already become accustomed to the life of gangsters, making easy money necessary work, free to commit serious crimes and still not be punished for what he does. It is our responsibility to avoid children during our premature phase, where we are not prepared to educate them correctly because we are not yet mature.

Because we still need to learn many things that we do not understand deeply about life, the world and ourselves. For not having a profession that allows us to have the financial conditions to give our future family a fair social placement, quality education and health ... We are disqualified to generate children or form families.

8. POLITICALLY RESPONSIBLE

Dissatisfaction with our public managers is notorious, because in general they exercise their functions for their own benefit and not for those who gave them the benefit of the power that today makes them sovereign over millions of voters in need of a management that allows them to better way of life. But there is no point in just murmuring because of the bad politicians that corrupt themselves day after day and ruin our society like a disease. We have to take action, go to the confrontation and punish those responsible, demand our rights as good citizens that we are, do our own justice against those who abuse our patience.

But how to act? What to do in the face of all this? Go to the streets in a direct confrontation with the authorities? Start a civil war? Threatening them publicly or just banging pots inside our apartments like a bunch of stupid pacifists who think this is enough to change the course of a nation seized by corruption?

We are politically responsible for them being up there, sitting in their places of power, dictating their own rules and governing us as they please. Popular revenge must be expressed at the polls at the time of the elections. There will be little point in making any kind of demonstration to show dissatisfaction with our government if we then exchange our vows for a plate of soup, a certain amount of benefits that soon dissolve, a public work that dissolves after a few months because it was done poorly.

The Brazilian population is one of the most stupid in the world when it comes to giving in to political appeals made during election campaigns for believing in the false promises that are presented by corrupt candidates who recently stole public coffers, diverted funds, were recently involved in many clearly confirmed scandals.

Hundreds of people or even thousands of them are opposed to the current Brazilian government and all the congressional benches that today destroy our hopes for a better future for us and our children, however, they will certainly vote in favor of his reelection because he will come with that old one angry talk that he has good plans for the country and needs just a little more time to be able to carry them out and will win the votes he needs to continue governing.

This always happens in relation to all the political strata of this poor nation, they are reelected at each new election at the expense of popular idiocy, especially the youngest people who understand nothing about public management, conforming to empty ideologies, unfounded promises and apparent innovations.

That in the end never leave the paper. This, however, does not happen in most countries around the world. Recently, in South America, the struggle of the Venezuelans against the dictatorship that oppressed them was seen on the news. Argentines, Chileans and other South American nations have also shown strength against corrupt rulers long ago. In the USA it was no different and in the last election they kicked Donald Trump in the ass for realizing the mistake of placing him as the government of the greatest economic power on earth.

Since he was a lunatic who did not clearly see a hand in front of his nose, creating public policies that were of no use to the North American population, a madman who challenged progress and order. Useless for social and world good due to its absurd way of governing. We Brazilians must act in the same way and not allow people without the least political maturity or respect for human life to take over the power of our country, denying them the vote on election day, giving other candidates the opportunity to govern and show better results as managers.

9. WE ARE ETERNAL BEINGS

Human spirituality, too, is a vitally important factor for there to be peace and progress in this world, since it depends on the action of divine power among men. I understand that skepticism about the existence of a Supreme Being capable of directing our steps and leading us towards the future exists, but that does not mean that this disbelief is true in its concepts. Regardless of whether or not we believe in the truth of biblical or religious information, nothing can deny that there is a greater power in the Universe.

Something supernatural that makes its own rules and we depend on them to progress or regress during our existence on this planet. Science tries to discover this enigma and as it fails to deny the idea of a spiritual world, completely real, beyond what is within the reach of our carnal eyes.

The man considers himself modern, evolved, wise, but he cannot accept the fact that his existence is not limited only to this small space of time where he is at this moment, but that we are spiritually immortal. The Holy Scriptures state that by creating man his image and likeness God, this Supreme Being, gave him immortality. However, because he chose the path of disobedience, he allowed sin to enter this world and with that he received the death of his physical body as a punishment.

Despite this, science, which insists on stating that we came from the evolution of microorganisms and the monkey, persists in discrediting the biblical reports, refuting the idea of being formed by body, soul and spirit. In this way, limiting ourselves to a mediocre existence where we will pass through this world in a brief birth, growth and a certain death without any spiritual evolution. However, there are those who believe what the Scriptures affirm to us and trust that physical death is only a gateway.

The spiritist doctrine of reincarnation has already been widely studied and even certain scientific branches have come to believe in its veracity, in fact they consider it even possible that human beings are not formed only of matter, but that our essence is made of something beyond what we can see and touch, leaving only the most materialists who continue to doubt this truth. The theory that we are a spiritual being within a physical body limited to exist only for a certain time, then age and die, leading us to have to abandon it, but remaining alive, is widely discussed.

Renowned scholars of different lines of thought argue among themselves if this can really be possible and strive to find an answer that satisfies such curiosity, resolving once and for all the doubt as to who we truly are, what we are essentially formed from, where we came and where we will go after death.

Who we are? — About the Mysteries of Our Existence

To think that our mission in this world is only to exist on earth for a short time seems a somewhat ridiculous role, taking into account the eternity that has the entire Universe. Considering the hypothesis of an infinite existence of the spirit is more comforting. But it is not so easy for the most unbelievers to accept biblical truths or to be able to agree with this possibility, for most human beings it is reserved for man to be born only once, to die and find himself in the most complete oblivion.

Worst of all, the Holy Bible gives us only the certainty that we are spiritual beings. In the Super Interesting Magazine published on August 9, 2017 we read an article on this topic. It gives us a glimpse of what we can understand as the reincarnation of the human spirit. There the author of the article tells us that The idea of a conscience that survives death and reincarnates in new bodies is almost as old as faith in divinities and arose independently in countless cultures.

According to his research, several cases of people who swore they had been the target of memories, visions and even went back in time through dreams and lived almost in an inexplicable reality their past lives that after extreme scientific and historical investigation were proven to be true. Individuals who lived around with revelations from a time hitherto unknown, never before visited or experienced in their current existence. However, after an exhaustive investigation, not only did they identify all events as real, but in fact it was proved that they existed at that historical time.

How to explain all this? Why do we seem to know places around the world where we have never been present? And the affinity that leads us to admire, like, love or hate and despise people right from the first contact without them even giving us reasons to detest it? The passion for a certain language, the dream of traveling to a certain country and knowing it at any cost?

The inexplicable visions where we find ourselves in the midst of so many unknown people, but that do us good? What can this be? Well, spiritists say that they are flashes and effects of past lives where we relate to different people and we were either friends or enemies, places where we walk or live, languages we speak, parts of the planet where we once lived. Truth or not what matters is that all this is an inexplicable mystery that still attracts the attention of thousands of people, researchers, scholars, religious people in the inner world.

And, even if the Holy Scriptures do not state that it is possible for the human spirit to reincarnate, there are some doubts. Yes, because Christ himself was reincarnated in a new and different body after three days of death. The prophet John writes at the beginning of his gospel that Jesus existed even before he was born of Mary, he lived in heaven with God and came down to earth, incarnated in human form and gave his life on the cross for us.

In the book of Genesis we find the story of a man by the name Enoch who was translated, that is, God took that man's spirit from his physical body and took him to heaven without having to pass through death. In the same way it happened to a certain prophet named Elias who, in the face of the perplexed look of his disciple Elisha, was taken by the Creator's hands and raised in the clouds without facing death with the guarantee that one day they would be born again of a woman and would honor the name of the Almighty with their lives. Jesus himself affirmed to his disciples that John the Baptist, the prophet, was the reincarnation of Elijah.

So if that was not reincarnation what would it have been? If those men were reincarnated, took on new physical forms and lived in the land of the living, why can't we, in this century, consider how small our faith is that we are the result of a sequence of existences in this world? King Solomon was described by God as the wisest man of his day and it was exactly he who wrote in his book Ecclesiastes.

Who we are? — About the Mysteries of Our Existence

The following sentence: "What it is, it's gone; and what is to be, has already been; and God asks for an account of what happened. "Ecclesiastes 3:15 Let us carefully observe the words of the wise man, for he makes it clear that the reincarnation is real not only of the human soul or spirit, but of everything that exists in this world. He implies that nature itself, animals and the infinite are renewed with a new garment from time to time, however, it remains the same essence of what one already existed.

Here we can understand why in the Old Testament God used the mouth of the prophets to rebuke the Israelites for their sins and claimed that they were continually stubborn since the time of the desert, after they left the land of Egypt. Now, but several generations have passed since Moses released about three million people from captivity. How could they be seen as his ancestors?

What we can witness here is that in the eyes of the Most High they continued to exist as their people even after physical death and returned to earth through the reincarnation of their disincarnated spirits centuries ago. It is clear, therefore, that the Creator gives the living man the opportunity to correct his past mistakes through each new existence received, but not everyone takes advantage of this rich opportunity. And this is the real reason for reincarnation to exist, it gives humanity the chance to put their feet on the tracks.

Fix your faults, mistakes, straighten your tortuous paths, evolve as beings endowed with intelligence so that in the future few or none of us will perish for all eternity. Yes, because it will not last forever, one day the Creator's patience with human rebellion will come to an end and then he will punish each according to his rebellions, pride and lack of fear. If we were all able to assimilate the enormous importance of this, we might change our attitude and prosper widely. Eternity awaits us, whether we believe in the essence of the spirit, in reincarnation.

Iin the succession of existences on this planet. But, the question that does not want to be silent, is: Where will we go when leaving this world? If in fact all this speculation about an afterlife is true, how will we live for all eternity, in the light or in the dark, with God or without him?

And even though this God is a myth, but we really are spiritual beings who will leave this body and continue to exist in another dimension, how will we be treated by the Universe that formed us?

Perhaps none of us actually have these answers, but we can find them if we start to believe that our existence will become infinite and that we are not here by chance. Limiting our thinking to just seventy or even a hundred years on this planet and then disappearing seems pointless.

SECOND PART

03 - WE ARE HUMANLY COMMITTED

I. COMMITTED TO OUR FUTURE

Everything we practice or do will influence our own lives and those around us. Our words, our thoughts, our actions, what we believe, teach, defend, everything will have an impact on other people. Whether directly or indirectly, in a positive or negative way, for the growth or backwardness of the society in which we live as a whole.

Great changes were made in the world only by the action or vision of an individual who were able to change the way of thinking in his time. We can mention people who died centuries ago, but have left their mark on history for their better evolution or even their almost total destruction. Names like Jesus Christ who in the first century of the Christian Era revolutionized the way of seeing and believing in an invisible, infinitely powerful and creator of the heavens, the earth, as well as the entire Universe.

This belief until then defended only by the Jews. The so-called "fathers of the church", like Peter, John, James and Paul who expanded the Gospel both in Jerusalem and throughout Asia. The 14th century reformers who gave their lives to allow ordinary people and not just Catholic clergy the right to read the Holy Scriptures in their own language and not in Latin. Great thinkers, idealists, inventors, philosophers, scientists changed the course of ancient history until it was possible to achieve a global evolution as we see today.

Because of the religious ideology preached by certain individuals, wars, fights, deaths, persecutions and incomparable conflicts have arisen. Political ambition, racism, the desire for power idealized by madmen like Adolf Hitler caused horrors like the First and Second World Wars. All of this shows us how our actions encourage those who observe us. Men and women passed through this world and left the mark of their steps wherever they went because they were able to show the crowd their particular way of thinking and seeing life, dissatisfied with their time, desirous of change. They did not agree with the current system that dominated their time, whether in the way of governing a nation, teaching the Christian faith, in the arts, in literature, in philosophical concepts ... There were always those who opposed customs and traditions.

Exposing their points of view, encouraging thousands of other individuals to also defend a radical change that would free them from the archaic thinking in which they lived submitted. We all have a duty to remain committed to the present in which we live, in order to shape it positively with our ideas, rejecting fables or archaic ideologies, opposing old policies, old concepts, backward views that only leave us stagnant, seeking the changes and transformations necessary for the future generation.

If we really want our tomorrow to be really different and bring us incredible news, we must stop the inertia in which we live trapped and start looking for radical social transformations that change the current view of acting and thinking of humanity, starting with ourselves and then with everyone those we can reach.

II. COMMITTED TO OUR LIVES

A society is formed by individuals who, grouped together, form families of different sizes, aspects, characteristics with particular and unique existence concepts.

Who we are? - About the Mysteries of Our Existence

However, despite their cultural, racial, religious or political differences, they end up having the same ideas for survival on this planet inhabited by billions of people who are willing to do anything not to lose the opportunity to remain in their physical bodies for as long as possible. The gift of living and existing in this world is something that has been given to us since when we were still generated in the womb and from then on our spirits.

Who have just been sent to a reality completely different from their previous state, they join the matter and feel the first contact with the hot or cold atmosphere of a place when they are born, which puts them in front of the possibility of exploring new and surprising things that, for their minds blank, it seems like a huge challenge.

From then on, the journey towards the unknown will begin in a process of growth in all areas and meanings, in a constant evolution of values not yet discovered. And it is during this advance in your existence, in discovering innumerable novelties, exploring each vastness of possibilities that each individual needs to be extremely careful not to end up following tortuous paths that will later lead you to perdition. Here the term to lose oneself has nothing to do with the condemnation of the spirit, but with oneself.

I often say in my lectures on the direction in which humanity has followed in the last centuries that if today society is enormously lost in its moral or spiritual concepts, it is because the new generations have failed to take into account that each path leads in the opposite direction to another .

In other words, if we are not careful to see where we are going, we may trip over a stone or fall into a precipice, whether literally or figuratively, dealing with the inevitable future damage that will arise in our lives as a whole, be it family, professional, social or any other area in which we will be inserted.

A criminal was not born committing crimes, a prostitute was born virgin and pure, a pedophile was also a helpless child before he grew up and made his victims ... critical situations to which they were subjected during their growth, while they matured and learned under harsh conditions an erroneous way of living and relating to their fellow men.

A life of material deprivation, constant poverty, can lead someone at any stage of their life to become a delinquent and start stealing, killing, taking possession of others' assets. An abusive childhood on the part of parents, relatives or strangers leads a man to commit rape, to rape children, women and other similar crimes.

For a more just, peaceful and balanced society in the future it is necessary that today we are all seriously committed to the quality of life that we have and offer our children, so that they can be good people and teach this to the generations that arise later their.

If the world population continues as it is now, defending pornography, homosexuality, sexual or social violence, an incorrect policy based on corruption, nothing will change tomorrow.

If our authorities continue to take away the right of parents to correct their children, to teach them severely the right way to go, if our young people and adolescents continue to practice their delinquencies without suffering for it, surely future generations will be worse than the current.

III. COMMITTED TO OUR DESTINATIONS

At the beginning of this work I wrote a topic about the fact that we are all what we think we are and this truth must be taken very seriously because our thoughts really influence our current existential reality.

Who we are? - About the Mysteries of Our Existence

If an individual firmly believes that he is a criminal or any other form of delinquent, that will be his identity before the Universe and he will respond to the height of what was idealized. In other words: The world around us attracts our daily thoughts and materializes them into something real for ourselves.

So, starting from this principle, answer yourself (a) who you really believe to be at this moment: You feel like being a person positively confident in your personal dreams, able to chase your ideals at any cost, willing to overcome all obstacles that appear to you along the way without hesitation or discouragement, the kind that even in the face of the worst difficulties you remain confident that everything will work out in the end? Do you think you are morally responsible with your actions, sincere, true, committed to the truth or the opposite of all of that?

If we are what we think, it is because God or the Universe, as he wants to call it, usually turns our personal beliefs into something concrete, real, true in our lives. Negative people live looking backwards or downwards and never move. Everything they set out to do ends up going wrong, they stumble in their own shadow, and their efforts to achieve purposes end in dire failures.

This is because the negativity of their thoughts takes away their faith in what they are looking for and without this essential condition for human growth we are unable to accomplish anything. We need to believe in the idea that a supernatural power directs us in this world towards our purposes and that it accepts as a rule what our minds define.

If we want a better future, then let us have confidence in this. If we want our destinies to be promising, let us think about it with more conviction, determining that this ideal be realized in our lives and that mysterious force that usually considers our desires will fulfill them.

Are we what we think we are? So let's think about being the best, bright, intelligent, important people in the face of the society we live in and the whole world, even if someone thinks otherwise about us. When I started to write and publish my first works on a digital platform in E-book format.

I was by no means a well-known writer, I was a mere beginner and independent because I did not have the financial means to publish my books through a physical publisher. But I confess that in my heart I was sure that one day the whole world would get to know and read my work, what I worked hard and produced in a cramped room in my humble home. Today, after almost six years, my works are sold by several countries in Europe, in the USA and the United Kingdom, arriving in China.

This happened because during those many nights of sleep that I lost hunched over the desk and keyboard of an old computer, I always believed in my potential, in the gift of creating, of writing, of knowing how to put on paper what I really think. In no way did I doubt that I was born to be a writer and that this would be my path in life, through him I would become someone recognized.

If many fail during the journey they live on earth, it is because they think small, do not believe that they can go further than they once were, stagnate and stop in time feeling like slugs and not as people capable of achieving the impossible. God in creating man said with all certainty and conviction that he (we) would be his image and likeness. We must believe that! In the preface to the self-help work entitled UNLOCK THE POWER OF YOUR MIND, by writer Michael Arruda, we read:

"You went through several books until you found this one. He looked at the cover, the title caught his eye, started to read the first sentence and decided to continue. However, I ask: did you decide each of these steps?

Who we are? - About the Mysteries of Our Existence

You may believe so, but the truth is that everything happened so fast that your actions were already decided before you could think about them, taken over by a deeper part of your mind: the subconscious, responsible for what we are and We do. What other paths are you taking in your mind without it consulting you? In his first book, Michael Arruda, president of OMNI Brasil, will show you how to take control of your mind and, consequently, of your personal and professional life. For this, he will present you the process that allows you to access your subconscious, identify the causes of pain and dissatisfaction and solve them quickly and effectively: hypnotherapy.

On this journey through your mind, you will learn:

— How the three parts of your mind work and what are the mechanisms that lead you to make your decisions, from the simplest ones to those that can transform your life radically;

— Why you are the way you are - and what is the root of your biggest problems and challenges;

— How to stop self-sabotaging yourself with habits, attitudes and beliefs that only take you away from your goals;

— How to become the captain of your life and never feel hostage to fate again."

The most distinguished and respected scholars of the functionalities of the human mind have already marveled at its ability to profoundly influence our lives positively or negatively. However, there are still few who really consider this fantastic discovery about the power it has to materialize our thoughts. I once heard someone say in a lecture that all he had achieved in the course of his life was to mentalize his dreams, the things he wanted to possess.

To be convinced that it is even possible that we take a few hours of the day or night to settle somewhere, in bed for example, close our eyes and think about something we want? Let's do this with faith, in the certainty that somehow the sovereign power that governs the universal guidelines will be listening to us and that it will make the impossible happen in our favor, removing all obstacles, uncertainties, doubts, impediments ...

And finally, we will see our hands touch what we imagine so firmly, with such insistence, with the certainty that our mental efforts would not be in vain. But, beware: Let us not make this a simple joke, we take it seriously because our success depends on it. Let us remember that Jesus Christ, whose wisdom transcends our limited imagination as human beings, more than two thousand years ago already taught that "everything is possible to those who believe". This statement by the Grand Master only confirms the power of our mind.

Let us always be optimistic, confident people, certain that we are also worthy of having a better life at our fingertips, that everything on this vast planet is rightfully ours, its endless wealth was created for humanity and we are part of it. If until today nothing has been important, nothing has been interesting, nothing has worked, so let's change this situation!

IV. COMMITTED TO OUR DREAMS AND IDEAS

We all crave something interesting in our lives, an ideal, a dream that at first seems even impossible to materialize, but that nothing prevents us from continuing to believe and persist in achieving it. There are, however, those who in the end end up conquering the fulfillment of those desires and live fully what they previously tried to achieve at high costs and who despite so many efforts never reach their goals. This is because some are more dedicated than others during the walk.

Who we are? - About the Mysteries of Our Existence

If someone is really committed in body, soul and wholeheartedly to conquer his space through society, to grow financially, to finish college, to possess material goods or to occupy a prominent place in this world, he will certainly succeed, but if he only wants in his does nothing to make that dream or ideal materialize will not be successful at all. There are several cases in which people who were once born and lived for a long time in the most complete misery today have become celebrities.

There are also reports of those who were born, grew up and lived a large part of their lives in a golden cradle and that are now in the gutter the margins of society that in the past honored them. We need to be entirely focused on what we want for our future and never give up on that until we finally achieve it. Let us remember that this world belongs to us, the planet earth is ours, it belongs to humanity, to each one of its inhabitants, mine and yours. Of course, apparently some are more deserved of this than others, but everything is ours.

The Universe or God listens and attends anyone who shows you some form of desire filled with the purest faith. He makes no distinction between white and black, rich and poor, men or women, young or old ... His answer will always be given by the size of our trust in him.

In the Gospels Jesus Christ tells us that everyone who knocks on the door will open it and whoever asks will receive it. This makes us clear that before the Divine Being or the force that governs the Universe there is no partiality of people, there is no such thing as deserving more than others or more or less favored.

If we ask and we do not receive it, it is because we ask little or without the necessary faith so that our petitions reach the ears of the Almighty or the most distant part of outer space and the power that governs the infinite can hear us.after all.

The greatest of all human beings who ever lived among men was the one who affirmed… that everything is possible for the one who believes and he did not immortalize himself by making up lies. Faith is the force that moves the hands of the author of Creation to answer our cries, we need to be more confident in that.

And if the reader is an atheist, oblivious to the existence of an extremely powerful God who intends to give him everything he desires then think of him as the action of the Universe that is above his head, but have faith and make him bless you.

Being truly committed to our dreams and ideals means never stopping walking, moving towards them with optimism and the certainty that we can reach them. Sometimes I think of names like Roberto Carlos, king of Brazilian popular music. Edison Arantes do Nascimento, Pelé king of football, and many others who came from such poor origins and reached the social top after ceaselessly seeking the fulfillment of their dreams and finally becoming recognized worldwide.

In the same way that they reached the highest point in their lives, all of us will also be able to go further and be graced by the response of those who hear us and wish to fulfill our requests. It is enough for us to dedicate ourselves a little more, to cry out with greater intensity not only with words, but with groans generated within our inner being, in the soul, in the spirit.

04 - ARE WE UNIQUE?

IS THERE ONLY SMART LIFE ON EARTH?

"The big question remains unanswered. Many believe it is difficult not to have a planet that harbors life anywhere. Lakdawalla believes that "if you look for places like Earth, you can find life or not; this will answer whether we are common in the universe or whether we are not. But life can also exist in ways that we don't imagine.

Scientists believe that Mars has harbored at least some type of microbial life in the distant past, and some even dare to say that there may have been other forms of life there in the past. Titan may have a life based on hydrocarbons like methane, which would be quite different from what we know. Scientists are also suspicious of Encilado, a Saturn moon that has an ocean of salt water beneath its frozen crust.

Going further, if we ever find signs of another habitable planet with a great possibility of harboring intelligent beings, we would come up with another problem: contact. With the technology currently available, sending a message can take centuries to reach an exoplanet, and any response would possibly take the same time to be received by us. Traveling personally to other star systems, then, is unthinkable with current technologies. "We would have to discover new physics," *says Quarles.*

"We would need to discover something totally new to fill this gap", whether for a journey of contact with other forms of life, or to colonize other planets. Star Trek, therefore, is still a reality that exists only in science fiction. Aliens may even exist, but we are unlikely to make contact with other forms of life in the near future. But if we have no way of communicating with other forms of life or traveling to planets around other stars.

Why insist on this incessant search for extraterrestrial life? Or even to study so many alien planets that we may never see up close? Lakdawalla explains, at least as far as planetary geologists are concerned: "Many of us study exoplanets just because it's cool. We only have eight planets in the Solar System, and now we find a wide variety of totally different planets out there," he says. So say NASA, with its list of planets of horrors released on Halloween - Source: Inverse, NASA

For hundreds of years man has been looking to the sky in search of whether he is the only one to live in the vastness of the infinite, from the study of the stars by the archaic astronomy of the Chaldean people to the most modern techniques used by modern science. part of human beings the burning desire to know more of what is hidden from their eyes in the infinite above their heads. If things were done with the naked eye before today, technology helps us to go beyond the clouds and take a close look at what's up there.

After the end of the space program in 2011, NASA spent ten years without pursuing its search for the existence of new intelligent lives in the Universe after the sad incident that occurred on January 28, 1986 with the CHALLENGER where all its crew members were killed. "The Challenger tragedy was a traumatic milestone for the American population and for NASA itself. It was on January 28, 1986, during the tenth mission (STS-51-L), 73 seconds after the launch of the space shuttle. Victiming all 7 crew members on board, including Professor Christa McAuliffe.

The timing of the astronauts' death is uncertain, but it was estimated that they died due to the impact of the capsule against the water. After 5 months of research, a commission found that the solid fuel rocket on the right side of the bus was defective: the sealing rings did not expand as they should and the gases escaped from the internal compartment. Later, it was discovered that there were many more problems than that. Involving some NASA decisions to launch at an inappropriate time, despite warnings from some experts involved in the subject. It was not possible to determine the exact cause of the death of the crew.

Who we are? - About the Mysteries of Our Existence

Although the recovery of the wreckage of the crew compartment offered relevant information, the results were inconclusive. Apparently, they survived the forces of the explosion and the rupture of the compartment, and probably lost consciousness seconds later due to the loss of pressure from the crew module. "

However, despite this horrible disaster and the end of the space research program in 2015, a new program would begin that would prepare young people aged 16 to 18 to become future astronauts who would be sent to Mars in the year 2030 according to the publication of the Magazine SEE done two years before.

"Sending man to Mars by the 2030s is NASA's priority. The agency must direct all of its financial resources towards the mission of overcoming technological and knowledge flaws that still make it impossible to achieve this advance in space exploration. That was what Charles Bolden, a veteran astronaut and current administrator of the American space agency, said at a conference held at George Washington University, this Monday, May 7, 2013 ". - Source: VEJA

What is certain is that modern man, like his ancestors, will never fail to investigate the infinite in search of answers as to the existence or not of other intelligent lives in the Universe, this doubt consumes the human spirit inside as if it were a cancer, the thirst for knowledge of what we do not understand is much greater than we can imagine and that is what drives us towards new technological and scientific discoveries that cooperate in the evolution of our race.

Many even wonder what we have gained with so many billions of dollars invested in space research, what return do we get from the studies done by NASA and the American government. New technologies! In addition to immense progress in science in general, the journey of man to the moon has made us evolve in several other areas since then, as many new techniques need to be created so that these trips through space can be made.

In the same way it will happen after having managed to go to Mars in 2030, we will certainly stop seeing the world and the life of the human race as we have seen until now, countless changes will happen in the technological, scientific, social area and a huge expansion in knowledge as a whole.

The new NASA program that began preparing young people and teenagers aged 16 to 18 to become future astronauts and who will be sent on the journey to Mars in nine years time will be one of the great achievements already accomplished in recent decades, as it will elevate considerably the human knowledge about the unknown and will make us go beyond the space boundaries. Upon stepping on the red planet's soil those fourteen astronauts will become world famous and the USA the first nation to send people for a unique achievement.

The question initially asked in this book about who we really are will finally find a conclusive answer, as we will understand that we are fantastic, universal beings, capable of going beyond the imaginary. We are fearless, pioneering, bold, determined, committed to new discoveries, tireless in always looking for new answers and if in fact there is intelligent life in other parts of the Universe we will discover why if they do not come to us, we will go to them.

05-ONLY EVOLUTIONARY BEINGS

1. THE IMPORTANCE OF TECHNOLOGICAL DEVELOPMENTS

We constantly evolve in our way of being, acting and thinking. Our actions and attitudes change over the days, months, years, this process of natural transformation is called physical and mental maturation.

As a child we think like boys, when young we behave as such, after adults we completely change the way we see life, the world around us and learn to choose our paths. The human being is not just a living being, but rational, intelligent, in an evolutionary process.

Because of this unique quality we differentiate ourselves from other beings that exist on earth and occupy the highest part of the food chain, we live in societies subdivided into several layers ranging from the lowest to the highest where each one occupies its role in favor to benefit others.

The poorest need to tirelessly seek their daily livelihood, so they render their services to the richest. The most powerful govern the laws, determine the social guidelines that often benefit the population or not.

There are the most intellectual, the geniuses of science and technology that favor us by creating and inventing new means of communication, treatments and cures for diseases previously considered incurable. They improve the world in a variety of ways and raise the bar for human knowledge. Anyway, we are beings that never stop growing in all areas of life that were once donated to us.

Unfortunately, also to a small number of individuals who have stopped in time and contribute nothing to this evolution, but it is a minority.

"One of the main drivers of the advancement of science is human curiosity, uncompromised with concrete results and free from any kind of tutelage or guidance. Scientific production driven simply by this curiosity has been able to open new frontiers of knowledge, to make us wiser and, in the long run, to generate value and more quality of life for human beings. Through its methods and instruments, science allows us to analyze the world around us and see beyond what the eye can see. The scientific and technological enterprise of the human being throughout its history is undoubtedly the main responsible for everything that humanity has built up to now.

His achievements are present from the realm of fire to the immense potential derived from modern information science, including the domestication of animals, the emergence of modern agriculture and industry and, of course, the spectacular improvement in the quality of life of all humanity in the world. last century. In addition to human curiosity, another very important engine of scientific advancement is the solution of problems that afflict humanity. Living longer and healthier, working less and having more time available for leisure, reducing the distances that separate us from other human beings - whether through more communication channels or better means of transport - are some of the challenges and human aspirations to which, for centuries, science and technology have contributed.

A person born in the late 18th century would most likely die before reaching the age of 40. Someone born today in a developed country is expected to live more than 80 years and, although inequality is high, even in the poorest countries of sub-Saharan Africa, life expectancy today is more than 50 years. Science and technology are the key factors in explaining the reduction in mortality from various diseases, such as infectious diseases, for example, and the consequent increase in human beings' longevity ". — Published on 11/07/2019 Last modified on 12/23/2020 at 1:45 pm — The Team of the Center for Research in Science, Technology and Society

2. THE IMPORTANCE OF SCIENTIFIC EVOLUTION

All this technological advancement allows the discovery of new scientific methods that in turn leads humanity to be more and more able to protect itself from biological threats, previously deadly diseases, increasing our life on earth, enabling man to carry out new research that it will lead our race to multiply their knowledge of what is perhaps still hidden in front of their eyes.

And this is, without a doubt, our greatest quality, because we are tireless in the pursuit of knowledge, we are curious, dissatisfied Science has not yet found a cure for chronic illnesses such as cancer, AIDS, diabetes and others.

But it has already discovered drugs that help control its harmful effects and allow its carriers to prolong their life without having to suffer from it.. Medicine evolves along with new scientific discoveries because one complements the other and in part relies on the help of technology that is seen as the mother of evolution.

Scientists spread across the four corners of this planet are tirelessly studying new possibilities. This joint effort benefits the entire human race because new ways of minimizing the lack of health, education, and the social as a whole appear at every moment. Before those who contracted the HIV virus received their death sentence immediately.

However, today there are drugs that control the viral attack and give the host a chance to live longer. It also occurs in relation to many other incurable diseases, where the patient is certainly at high risk of life by contracting them.

However, being treated correctly with the right drugs can prolong its existence in this world and even survive the deadly attack of viruses and bacteria. Medicine hand in hand with science and the two allies with technological expansion become unbeatable.

Abdenal Carvalho

"For those of us who developed our scientific task in the second half of the twentieth century, the balance in the relationships between individual, science and society ended up shaken by changes of great importance in the ethical consideration of scientific activity. In view of the great advances achieved, the extraordinary increase in scientific and technological knowledge and its bibliographic collection, and the speed and reach achieved in the dissemination of such advances, those changes in the relevance of ethical aspects have passed almost unnoticed by many.

Humanity was shaken by events that were almost inconceivable, and certainly incredible, that made it aware of such an ethical dimension. Of the events that occurred in the decade of the 40s that marked the change, are: the unprecedented attitude of the Nazis of massive extermination and disgusting experiments on humans, and the atomic bombing of civilian populations carried out by the allies.

But there was much more. Biotechnology and molecular genetics opened up frontiers and possibilities that were previously unthinkable. The prospects for radical changes in the distribution of species and their control, including for humans, overcame the concern of a possible misuse of the results of specific investigations. Outside the scope of laboratories, the progressive and irreversible destruction of nature and its resources, the discoveries of climate change that point to an uncertain future for life as we know it and the restrictions imposed by market protection mechanisms, are new scenarios that awaken concern for its ethical facets ". — *INCI v.31 n.7* ***Caracas jul. 2006***

Brazil is still crawling in terms of scientific and technological progress because our governments show little or no interest in investing in these areas, the little knowledge we have and make use of comes from abroad. From other countries like the USA and more recently from China, Who would have thought that that once poor and discredited nation would become the second largest world power in economics and technology and even lend us its knowledge. Although it was the chinese scientists responsible for the viral devastation that currently destroys...

Who we are? - About the Mysteries of Our Existence

… The lives of millions of people around the planet, it was they who helped Brazil in selling respirators and masks to try to prevent death or contamination through Covid-19 , because here we didn't have enough.

Now imagine a nation with more than two hundred million inhabitants without the minimum protection needed because their governments do not give a damn about the realization of new research in the area of science and technology, here everything has to come from outside so that it is possible corrupt politicians to embezzle public money. Unfortunately, we are still horribly backward in relation to other countries, but we can only hope that one day it will change the chaotic situation in which we find ourselves.

As we crawl through scientific and technological evolution the rest of the world is creeping forward to the highest peak on the mountain of knowledge. Most Brazilians are already born with the mental laziness syndrome, our people do not dedicate themselves to reading, are addicted to social networks, games, and other similar things, but they hate to seek knowledge that will transform their lives and that of the entire country.

The truth is that a young man born in America or in any other part of the evolved world knows more than hundreds of Brazilians. We are seen out there as an extremely backward people and that is still true. Brazil is the country where books are least sold, regardless of genre. Many bookstores seen as important have already closed their doors because there was no clientele.

Sales were very low and it was no longer possible to remain active in the literary market. As a writer, I know this reality a lot and if today I live exclusively on the sale of books, it is because I opted for independent publishing, through large platforms such as **BLURB** and **INGRAM** that distribute my works globally, translated into two more languages.

Whoever chooses to wait for the sales of his literary creations to happen through physical bookstores will see them morph on the shelves. Our leaders despise knowledge, cultural, scientific and technological development. The people follow their example. I believe that if future generations start to read more and encourage their descendants to have a passion for the search for wisdom, our nation will evolve in all areas and one day the Brazilian people will no longer be taxed as lazy and disinterested, they will no longer depend on the discoveries and foreign inventions because it will be possible for them to accomplish great things in all areas of human knowledge.

But, for this to really happen, we need to change our concepts about the need to evolve, to grow, to learn ... A nation will not have any chance to evolve and prosper if it does not invest in research, discover new paths through culture, education, and interact with new world trends. Brazil, as well as all the other countries in Latin America, remain totally trapped by a constant inertia in relation to global growth.

Often because they refuse to invest resources in the area of studies and research or because they do not have the economic means to do so. , but Brazil has this condition and refuses to carry it out. Our country finds itself in decades of scientific backwardness, not because of a lack of resources, but because of the sheer neglect of our leaders. Who instead of thinking about growth in these areas, prefer to invest in other purposes, almost always without short-term return, without taking into account diversion of public funds through the vicious channel of political corruption.

3. CULTURAL EVOLUTION

A people or nation differs from others when it demonstrates greater scientific, technological, cultural and economic knowledge.

Who we are? - About the Mysteries of Our Existence

The United States stood out more than all the countries in the world for almost two centuries after the Second World War, since after that sad episode it remained as the only economically structured nation, occupying the number one position in the global financial elite. But over the course of several decades other nations have also evolved and today China is the second largest power.

However, they wisely refused to remain stagnant only in their economic growth and began to invest in scientific, technological, space research that led man to step on the moon and thus occupied the highest position on the planet's cultural scene. Now, to further increase its importance, it intends to send the first human beings to Mars and in doing so it will become unbeatable. Certainly, competitor China will want to do the same, as it did with Russia in the past.

Unfortunately, third world countries like Brazil and others in Latin America are still crawling in several aspects, mainly in the cultural economic area. We remain poor and lacking in knowledge because they treat us like irrational beings and we accept this naturally.

Our government officials refuse to invest in education and we agree with that by going to the polls to vote for them whenever there is a new election. The Brazilian people got used to living under the weight of ignorance and neglect, since our mediocre origin was like that. Coming from peoples like the Portuguese who knew very well how to exploit the most ignorant, the illiterate and superstitious Indian.

The blacks without the least education and given to the habit of obeying without complaining, we are too passive, we refuse to fight for our rights. Nothing is more sad and shameful than to deny a nation its constitutional right to evolve culturally, because only then will it achieve the growth it needs to stand out among other peoples and give its inhabitants a better quality of life.

First world countries shine on the world stage due to their many riches acquired over the years, but these financial resources did not fall from the sky, they struggled with all their strength to acquire them and one of the weapons used was exactly the cultural development by which they found ways to evolve.

How can we move towards a better existence in this world if we find ourselves with our feet and hands tied in ignorance and lack of knowledge? How can Brazil or any other country develop in other areas, move in other directions as a nation if its inhabitants are denied the right to educate themselves, learn new things, achieve greater knowledge that will lead them to see further? It is sad and even shameful to admit, but most Brazilians despise the habit of reading and studying.

Perhaps the reader thinks that I have reserved this topic to criticize my people, to diminish the cultural value of my country, but it is nothing like that. What I do here is to try to awaken this nation to the importance of dedicating itself to knowledge and making it the main factor of our lives, because without wisdom we will always walk in the dark, blind as moles, hidden inside our burrows conformed with crumbs. But do not think that this occurs only here, several other peoples suffer the same or even greater neglect.

We need to react, demand our right to a quality education for our children so that our future generations can be part of more evolved groups and not follow the example of their ancestors who remained in the cultural blindness, on the margins of society as useless and without no value at all.

06. WHO ARE WE REALLY?

I. WE ARE THE FUTURE OF THE HUMAN RACE

I don't know how the reader has ever stopped to think about his true importance as a person in this world, if he has ever looked in the mirror of his conscience and asked why he is here on earth, what is the purpose of his existence on this immense planet. I sincerely hope so, because whoever has a minimum of intelligence stops and thinks about these things, after all, reflecting on our lives is not only a philosophical act, but necessary for our evolution as a human being.

Among so many other importance that we have for the world around us, we must consider the fact that we are the future of the next generations, what they will be depends on who we are in this present where we are.

Our descendants will be the exact reflection of what we are at this moment and in the course of our journey through the labyrinths of human existence, it is useless to wish that they will be evolved and with remarkable characteristics if today we remain sitting on the edge of the path of ignorance.

If we dedicate ourselves to grow in knowledge, become wise in relation to the mysteries of life and the Universe as beings that did not conform to the inertia that would lead us to remain stagnant in the condition of a people or nation whose tradition of the country where culture was born is despised, so those who descend from us in the future will be and will be part of a special class of richly evolved human beings. We need to shine our inner light, rise to the highest point of life, be important to set an example for them. We are the draft of the future of the human race, through us it is being written at this moment.

We are like the pages of a huge book where the Universe describes tomorrow, both good and bad. If someone today is bad, ruthless, unfair, cowardly, liar, if everything about this individual is obscure and negative, then those who descend from him in the future will be almost entirely in his totality. But if there is only goodness, wisdom, intelligence, humility in us, if we are enlightened by the light of God, so will our descendants.

If the modern world has turned into chaos, it is because past generations have given little importance to this reality. If people kill, steal, exploit the poorest, they corrupt themselves because they have such characteristics in their DNA. Individuals who love the practice of evil will generate children, grandchildren and great-grandchildren who will be able to socially express the same moral defects. Jesus Christ said in one of his lectures before the crowds that a bad tree cannot bear good fruit.

This wise placement of the Messiah is the purest truth, since people of bad character pass through their acts a certain negative incentive for those who assist him, often leading them to copy their injustices. Of course, a certain person can present a positive and morally approved personality even though he is born to parents whose way of proceeding is reprehensible.

But in almost all cases children tend to follow the same example as those who served as guides. But, even in the face of this possibility and the imperfection of the modern world in which we live, of uncontrolled violence, of a youth lost in drugs and prostitution, social injustices, pandemics that lead death to put an end to millions of lives, of all this darkness that around, we need to remember who we really are. We are the essence of life, of nature, of this planet and what is most important in the Universe, since despite the speculations nothing proves concretely that there are other intelligent lives elsewhere in the cosmos.

Who we are? - About the Mysteries of Our Existence

At least for now, the human race is still at the top of the galaxy's sidereal, universal existence. We are the only ones to have rationality, a characteristic above other living beings and if there is an extraterrestrial up there with intelligence superior to ours, then appear.

We are the ones who were born, grew and multiplied, giving rise to new generations. We are the ones who learn to love, smile, cry, create our own dreams. We are the ones who struggle with all our strength in the conquest of our ideals, in search of being better than we were before, who created spaceships and went up to space in search of other planets, of other lives, to be able to understand where we came from and where do we go after this life cycle falls apart.

We are a persistent race that never gives up creating and inventing new ideologies, who research and explain the unknown, we find cures for our own diseases, we carry the whole world in the palm of our hand through a technology that we invented ourselves, we learned to fly, to speaking several languages, we went to the moon and planted our mark there to make it very clear to such extraterrestrials that even though they are made of flesh, bones and blood there is a force within us that motivates us to never stop walking.

We must never be intimidated and think that we are nothing in this vast Universe of countless surprises and news, humanity is special, it is eccentric, courageous and knows how to take its place. Our history of victories and failures does not reduce or diminish our importance, but, on the contrary, only makes us more mature to move forward in the process of evolution.

We are physically fragile, yet we endure the heat of the most intense passions, struggles, disappointments, wars against the worst enemies, pandemics, almost incurable diseases and if many of us stay on the path others go on without ever getting discouraged.

The species emerged about 350,000 years ago in eastern Africa and acquired modern behavior about 50,000 years ago. However, archaeological evidence published in 2017 suggests that it may have spread across the African continent some 300,000 years ago.

We are like a seed planted in fertile soil, always multiplying and generating new fruits, giving rise to new lives that in turn will also branch out. Good and bad fruits are born from each new plant, some are used for our food and others are poisonous. Likewise, we are human beings, because in this world there are people of good or bad nature, who build or destroy the planet, but we still have our importance as a living being endowed with reasoning.

Life is different for each individual, while some live only a few years in this world, others come to exist for more than a century. Science and medicine point to physical and hereditary characteristics as the main factor for certain individuals to remain in their physical bodies for more or less time. There are those who are born and die shortly thereafter, those who last a few decades and those who continue until they tire of their own existence. The Holy Scriptures, on the other hand, say that the final decision is from God, the Creator of life.

David wrote in his Psalm 129 that life belongs only to God and it is up to him alone to decide who lives and who dies. The Lord of the entire Universe knows the exact amount of our days on earth. Well, whether or not I believe that we were created and placed in this world with a greater purpose than just to exist as a living soul, to grow and multiply.

I understand that humanity was and continues to be placed on this vast planet with a specific purpose beyond procreation. In my personal opinion, being born just to generate new lives and thus forming new generations is too mediocre a thought for a project of such greatness as we are all.

Who we are? - About the Mysteries of Our Existence

Generated from God or the Universe at a time when certainly important ideas were formed about us . We are, yes, parts of a great project designed for the growth of this planet, for the development of the natural riches that exist in it, for our own evolution, to become beings of vital value in the eyes of the one who gave us life.

For this reason, I cannot see myself and others only as an unfinished draft, without any expression, without a definite purpose, nothing more than simple people. And it is by looking at those people who have decisively changed the world that I affirm that thought. We are part of a universal plan that will profoundly transform things that are still far from our eyes.

Hidden from our limited understanding, but that exist and will one day be clearly shown to future generations. Today represents tomorrow's platform, it is the map that the human race draws today that will later guide it towards the final purpose of those who created it, we are the hope of those who are yet to come. That is why we all have a duty to seek to accomplish great things that change our current history, who are we really? We are the past, present and future. An image of what exists beyond the horizon.

Heyond the hills of the imagination, well beyond where our carnal eyes can see. You and me, dear reader, are not just here to add up to billions of other people. Our brief passage in this world has a greater meaning than just being among so many other human beings, we were sent to make a difference.

"Making a difference and a normal attitude in people, because each of us wants to somehow leave the mark of his performance on the record of his competence, showing how much he can contribute in a given situation. The people around us simply love that each one of us makes a difference, whether they are our coworkers, boss, partners or partners, family, friends, etc, make a difference and leave the commonplace.

And do differently, and give the best of us; when we do not want or can make a difference, when we feel unmotivated, helpless, when there is a stance of whatever, when we feel victims, there is certainly something wrong, and as if there was a disease. Make a difference and positively surprise people, doing something more that was not expected, and somehow exceeding expectations. Making a difference means delighting people.

Creating that magical environment in which people can say: for me, in that moment, in that place, you made a difference! We instantly recognize someone who makes a difference: it can be a salesman in the store, a waiter, a bus collector, a parking guard, a co-worker, an inspiring leader." — **Gustavo G. Boog and Consultant and Organizational Therapist, conducts projects to increase personal, group and business competence.**

07. WE ARE TIME TRAVELERS

The shortness of life that the human race has shows that their stay here on earth is just a quick passage and what defines the maximum time of each existence is the extension of the journey from the birth of a person to the arrival at its main objective determined by the Creator or described by the power of the Universe (as the reader wants to name the force that created us and sent us to this existential plane), which is in reality the mission to be completed by each one of us.

We can take as an example the passage in this world of several individuals who, like a comet, shone in artistic life and at the height of their career, left the other side still in their prime, leaving only the longing for the fans. Sometimes someone battles tirelessly to earn money, become a millionaire.

Stand out in the professional field, be famous, create something that gives him personal prestige, leave his mark on this world and barely knows that everything he wants to do has already been determined beforehand, written by the hands of fate.

He does not understand that as soon as he conquers his most ambitious dreams, he will end his journey, leave his physical body and cross the bridge that separates the two worlds where the living and the dead live. In reflecting on this, we need to be aware of the mistake we make whenever we are distressed because we have not yet reached the achievement of certain goals that we have proposed to achieve during our journey in this world.

Sometimes we even get angry, we blame this or that reason, we murmur against the heavens, we express hatred and resentment. But, as the wise King Solomon affirmed centuries ago, there is time for everything. His words were:

"Everything has its own time, and there is time for the whole purpose under the sky.

There is a time to be born, and a time to die; time to plant, and time to uproot what has been planted.

Time to kill, and time to heal; a time to tear down, and a time to build.

A time to cry, and a time to laugh; a time to mourn, and a time to dance.

A time to scatter stones, and a time to gather stones; time to hug, and time to get away from hugging.

Time to seek, and time to lose; a time to keep, and a time to throw away. Time to tear, and time to sew1.

A time to be silent, and a time to speak, a time to love, and a time to hate, a time of war, and a time of peace ".

If we set any goal and wish to complete it as soon as possible it is necessary before we think about the consequences of this risky decision, because the anguish that is consuming us from the inside will accelerate even more the wear and tear of our body, its main internal organs like the heart, brain, nervous system and attract death as soon as possible.

According to scientific research data, bad mood can even create stomach wounds, causing gastritis, ulcers and even cancer. Therefore, let us learn to wait patiently for the right moment of everything because this despair that usually overwhelms us when we wait for the completion of our expectations will not cause them to materialize overnight.

Who we are? - About the Mysteries of Our Existence

Human life is a journey that in some cases can be long or short, depending on the distance to be traveled by each individual in particular and the amount of luggage that is carried on the shoulders. I like to believe that nobody dies before the right time, I believe that our stories are complete because the Force that governs our existence on this planet is perfect.

The most religious believe that we came from the dust of the earth and were shaped by the Creator's own hands. Skeptics attribute our origin to the evolution of the species. I prefer to believe that we were created by Heavenly Father, by an immensely powerful, invisible, omniscient, omnipotent and omnipresent Being. I consider the sacred text found in the book of Genesis, which says:

"In the beginning God created heaven and earth. And the land was shapeless and empty; and there was darkness on the face of the deep; and the Spirit of God moved on the face of the waters. And God said, Let there be light; and there was light. And God saw that the light was good; and God made separation between light and darkness.

And God called the Day light; and in the darkness he called Night. And it was evening and morning, the first day. And God said, Let there be an expansion in the middle of the waters, and let there be a separation between waters and waters. And God made the expansion, and made a separation between the waters that were under the expansion and the waters that were over the expansion; and so it was. And God called the Heavens expansion, and the evening and the morning were the second day.

And God said, Gather the waters under the heavens in one place; and the dry portion appears; and so it was. And God called the dry land Earth; and to the gathering of the waters he called Seas; and saw God that it was good. And God said, Produce the land green grass, grass that gives seed, a fruit tree that bears fruit according to its kind, whose seed is in it on the earth; and so it was. And the earth produced grass, grass giving seed according to its kind, and the fruit tree, whose seed is in it according to its kind; and saw God that it was good.

And the evening and the morning were the third day. And God said, Let there be lights in the expansion of the heavens, so that there is a separation between day and night; and be them for signs and for certain times and for days and years.

And be for luminaries in the expansion of the heavens, to illuminate the earth; and so it was. And God made the two great luminaries: the greater luminary to govern the day, and the lesser luminary to govern the night; and made the stars. And God put them in the expansion of the heavens to illuminate the earth. And to govern day and night, and to separate light and darkness; and saw God that it was good. And the evening and the morning were the fourth day. And God said, Produce the waters abundantly reptiles with a living soul; and the birds fly over the face of the expanding skies.

And God created the great whales, and all the living-soul reptile that the waters abundantly produced according to their species; and every winged bird according to its kind; and saw God that it was good. And God blessed them, saying, Be fruitful and multiply, and fill the waters in the seas; and the birds multiply on the land.

And the evening and the morning were the fifth day. And God said, Produce the earth a living soul after its kind; cattle, and reptiles and beasts of the earth according to their species; and so it was. And God made the beasts of the earth after their kind, and the cattle after their kind, and all the reptiles of the earth after their kind; and saw God that it was good.

And God said, Let us make man in our image, according to our likeness; and rule over the fish of the sea, and over the birds of the sky, and over the cattle, and over all the land, and over every reptile that moves over the land. And God created man in his image; in the image of God he created him; man and woman created them. And God blessed them, and God said to them, Be fruitful and multiply, and fill the earth, and subdue it; and rule over the fish of the sea and the birds of the air, and over every animal that moves on the earth. And God said, "Behold, I have given you every herb that gives seed, which is on the face of all the earth; and every tree, in which there is fruit that will give seed, will be your food.

Who we are? - About the Mysteries of Our Existence

And to every animal on earth, and to every bird in the sky, and to every reptile on earth, in which there is a living soul, all green grass will be for food; and so it was. And God saw everything he had done, and, behold, it was very good; and the evening and the morning were the sixth day.

Thus the heavens, the earth, and all their army were finished. And when God had finished the work he had done on the seventh day, he rested on the seventh day of all his work, which he had done. And God blessed the seventh day, and sanctified it; because in him he rested from all his work that God had created and done.

These are the origins of heaven and earth, when they were created; on the day that the Lord God made the earth and the heavens. And the whole plant of the field that was not yet on the land. And all the grass from the field that had not yet sprouted; for the Lord God had not yet made it rain on the earth, and there was no man to till the earth. However, a steam rose from the earth, and watered the entire face of the earth. And the Lord God formed man from the dust of the earth, and breathed into his nostrils the breath of life; and man was made a living soul. And the Lord God planted a garden in Eden, on the east side; and he put there the man he had formed ".

This is the biblical theory about the origin of humanity, the heavens and the earth, as well as everything in them. And because I consider this to be the most true, I particularly defend the hypothesis that God is real and as David said it is his alone to intervene in our existence, determining when we are born, die and where we will go after we leave here.

For unbelievers as to the veracity of the sacred texts, the explanation of the origin of human life on earth is quite vague. And for centuries they have been trying to find an explanation that fits their trustworthy standards. But regardless of the belief of each one about our true origin what really matters is to remember that we are time travelers and he doesn't usually stop.

Even sleeping the hours, minutes and seconds pass and our physical bodies grow old, lose strength, follow the natural process of disincarnation of the soul and we gradually approach death. As my old father used to say, death is the only thing that we must certainly wait for regardless of social status or how important we may be, because at the exact time it will knock on our door.

And it is quite true, after all, he left a long time ago. Solomon addressed this issue by stating: "The same thing that happens to one will happen to everyone" In this trip that we take each one in its own way and pre-determined before we are even born, we go through several obstacles, crossroads, detours, we step on thorns, we create calluses on our feet, we scratch our knees and sometimes we shed excessive sweat to reach the end of the journey.

And this end means to achieve or not to achieve everything that we aim for in life, to realize or not to fulfill our dreams, goals and ideals, this is because not always what we seek is in fact according to what the destiny, God or the Universe determined. However, no one has the right to open their mouths to blaspheme and say that they have not fully lived their history here on earth, that they have not been given the same possibilities of conquest, that they have been denied opportunities, etc.

What happens is that certain people have extreme ambition and start wanting to have what was not given to them as a right, that is, the infinite power that governs our lives writes a certain story so that we live it down here, however, our eyes and heart they want something else. Hence, the result could not be anything other than the most complete disillusionment that has led many to commit suicide or to live in rebellion against everything and everyone. We need to understand as soon as possible that if human life is a story written by those who created us before we were even sent to this world.

Who we are? - About the Mysteries of Our Existence

It cannot be changed, we are not allowed to change the course of things, the direction we were assigned to follow. it will have to be fulfilled. Because some are born poor and end up rich and famous, the opposite is believed, but it is a mistake. If someone was born in the most complete misery and in the course of his life became an important person in society, he gathered enormous wealth.

Fame and prestige for himself, that does not mean that he changed his destiny. In fact, she simply followed the path that had previously been proposed. Walked exactly where her story described, mortal man is not allowed under any circumstances to modify the plans of destiny. He is the one who tells us what to do, how to live, where to arrive and as much as we fight against it, we can do nothing to prevent.

Our steps from taking us directly to where he chose. To think that we own our own lives is foolishness without size, we are governed, directed, governed by a greater force that is always in control of everything, dictating the rules, pointing out how far we can go. Sometimes we mistakenly say that a certain individual has overcome his disabilities, crossed his limits and reached beyond what he could. However, this is just a tremendous mistake. If a blind man has learned to play the piano, guitar, drums or any other musical instrument.

Becoming famous in what he does does not mean that he has exceeded his limits, but that he has reached where his destiny wanted. Many physically and visually impaired people have done fantastic things than others with complete perfection. But that does not change the concept that we only do what we were predestined, everything we do in life or achieve was already part of an architectural plan even before we were here. You and I will never be more than our existential stories allow, the author of life has already written our beginning and end, nothing can be changed. If we shine in front of the crowds.

If we swim in the rivers or seas of fame and money, if our name reaches the height of the stars, let us know right away that this was our destination. In no way do we change it or boldly rewrite our history. I again mention names like that of singer Michael Jackson who became the king of pop, an icon of American music that has won millions of fans around the world with his fantastic art of singing and dancing that is still unsurpassed today.

However, we know that before such success he had a sad and poor childhood. Did Michael change his own destiny and thus reach the top of fame? No, if we look closely at the trajectory of that celebrity, we will see that in the beginning it was his brothers who were successful, who demonstrated their talent for music while he was just a shy and repressed child. But after he had the chance to reveal himself in the artistic world, he showed his real value, even in his childhood he recorded songs even in our unforgettable days, overcame fear, shyness and went further.

He was determined to have glory as the lord of music and dance. Since his mother's womb his story of fame and success had already been written by the hands of the Creator of life and only him would this privilege be given within the Jackson family. Looking at the Holy Scriptures we can read several accounts about the lives of several heroes of the Bible who experienced similar situations in the past. It is the case of Joseph who was chosen among his brothers to become governor of Egypt, David who was crowned king of Israel, Solomon who became his successor.

Moses who was separated from his mother to be raised by the daughter of Pharaoh and in the distant future to free the Israelites from slavery. Men who lacked the luster to take on such important positions, but who were destined to fulfill such great missions among their people.

Who we are? - About the Mysteries of Our Existence

Regardless of looking at the examples described in the Holy Book or looking at the world around us, we will see countless cases of people apparently unworthy of reaching the podium of human existence. We were all born with our own history noted in the Creator's notebook of life, destiny, the strength that put us here on this immense planet. He has already predetermined where we would be born, whose children we would be, what gifts we would have, our professions, talents, opportunities and even the difficulties we would go through. Nothing happens to us by chance.

However, there are several ways of seeing and reflecting on this point of view, because both in philosophical and religious terms there are controversies about whether man is governed by the force of destiny or God, following in a straight line or going through certain deviations until reach the end point of your journey.

There are those who agree with this theory, but others disagree claiming the free will given to man by his Creator. But through the mouth of the prophet Isaiah, God Himself claims to have mercy on anyone who wants to have mercy.

Well, if he himself says that he chooses who to benefit from his infinite mercy then it is clear that the idea that there is a chosen people on earth whom he will allow to live with him in a future world and that others will not give such privilege is true . This, taking into account the Christian religious idea about predestination.

"Christians differ on one point or another, but in general they see Predestination as God's plan for humanity, which comes to pass when He chooses his elect. From the human point of view, every individual has the right to agree with the Creator or to repel him. Among these different views, the various religions give greater prominence to one or the other angle of this doctrine. This theory is also linked to materialistic, spiritualist, polytheistic groups, to karma, to the idea of destiny, to various philosophical beliefs and to various religions, under the conviction that if the future is practically programmed in advance.

So only a small number of events deviate from this pattern, while others believe in the preponderance of chance and luck. The person who makes a free choice can be based on an analysis related to the environment or not, and the choice that is made by the agent can result in actions to benefit it or not. The actions resulting from their decisions are subordinated only to the conscious will of the agent.

The expression free will usually have objective, subjective or paradoxical connotations. In the first case, the connotations indicate that the performance of an action (physical or mental) by a conscious agent is not completely conditioned by antecedent factors. In the second case, they indicate the point of view of the agent's perception that the action originated in his will. This perception is sometimes called "the experience of freedom. The existence of free will has been a central issue in the history of philosophy and more recently in the history of science".

Regardless of the type of belief that we may have about predestination, whether religious or philosophical, the truth is that life is a story and we are its characters, each with the mission of correctly representing its role. Whoever writes this play where we act as actors is still a very heated discussion among those considered most skeptical about divine existence, but Christians attribute to God to be the author of this human feat. It doesn't matter if God exists or not, someone wrote it.

If by chance what happens to us in the course of life was mere chance, if anyone could simply lead the way of their destiny, go in the direction they wanted, then we could all become prosperous, accumulate wealth, reach the podium and raise our trophies.

If free will is real, why are we penalized for rebelling against the dictates of destiny? Because in making our own choices do we suffer harsh consequences? Certainly someone will want to explain that if the outcome of our free will decisions is because we are not wise enough to make the right choices.

Who we are? - About the Mysteries of Our Existence

Then as people with little or no cultural education happen to reach the top while others rich in knowledge do not seem to come out. of the place? No, I still believe in the possibility of some force in the Universe guiding us towards a certain purpose. Human life is not and will never be absolute, vague, without control.

We were sent to earth to live a story written by a greater power, we received a mission to be fulfilled that will surely lead us to make a difference among the human race, whether good or bad. I think of good people, humanitarian, defenders of nature, doctors who save so many, scientists who find a cure for so many ills, create technological advances. I also consider those that cause so much harm to their fellow men, criminals of all kinds, corrupt, kidnappers, murderers.

Each one follows his destiny in a linear way, fulfills his goals, benefiting humanity or not. Good and evil coexist in the same system as light and darkness compete between day and night. If we look carefully, everything exists for a certain purpose, there is a definite purpose for everything in this world. It is as if the wicked were the expression of sin, of darkness, the dark and negative side of our being. And those who practice kindness were the reflection of our positive side, the light that we bring within.

The drop of love and sincerity that still remains in the modern man who lives lost amid so many contradictions. The earth is full of goodness and evil at the same time, on the one hand those who proclaim salvation and others who deny biblical truths about a true, loving and ready to forgive God. The science that invades space after other lives, death on our streets.

"Contradictions or paradoxes are thoughts or arguments that, although seemingly correct, present contradictory conclusions or consequences. Today's world is riddled with contradictions. Some of them are so cruel that they go beyond the semantic or logic fields.

Abdenal Carvalho

Since they directly interfere in the simple form of or human survival. In the light of this tragic picture, a simple glance at the general panorama of modern life, especially in urban life, is enough to see that our priorities are reversed; that is, we are all, or almost all, fighting for instrumental values, consumer goods. Whereas knowledge (as truth) moral values (as good); vital values (health and well-being) are relegated to the third level. " — **Francisco Assis dos Santos** —*17/05/2014* — *Gazeta do Acre. with*

We live in a confused society, totally lost in the pursuit of material values and forgotten about other concepts such as those that can lead us to understand more about who we are, why we are here, where we will go and what our real purposes are as human beings. The history of each one of us has already started.

And the majority is approaching the final goal without even initiating the fulfillment of their tasks demanded by the strength of the universe that placed us in this existential plane. I became a writer and through this craft I can reach people all over the world and talk to them about subjects like this, awaken in them the desire to get to know each other better, discover their true path in life, understand that their passage on this planet should not be in vain .

And you, dear friend, what have you been doing to change the way of thinking of your fellow men, how have you contributed to the evolution of your species, how have you been a participant in the changes of your time? Do not forget that we were born for a particular reason, everyone was given a purpose that will leverage greater knowledge, achievements, discoveries for future generations.

Christ, the greatest man that mankind has ever known, once told his followers that God gives men talents to multiply them and that when they appear before him without having properly completed this task they will be severely punished. What gifts or talents do we have? Are we putting them into practice or multiplying them?

Who we are? - About the Mysteries of Our Existence

What will be our explanation before the one who separated us so that we could do so? Will we say that we did not have enough time to exercise them or will we admit our negligence? The end of the journey for most of us is already approaching and what have we done so far? Are we part of a select group of people who have contributed to human evolution or do we stop in time with endless inertia?

For those less believers in what the Scriptures say, I ask you to consider at least the fact that some powerful force rules this world and that we did not fall to earth anyway, but were sent and with a certain purpose. Who brought us here? What do you want from us? Are we being useful to you? Let's think about it seriously!

"For this is also like a man who, going out of the land, called his servants, and gave them his goods. And to one he gave five talents, and to another two, and to another one, each according to his ability, and he was soon gone.

And when he was gone, the one who had received five talents negotiated with them, and gained five other talents. Likewise, the one who had received two, he also won two others. But the one who had received one went and dug in the ground and hid his master's money.

Long afterward the lord of those servants came and accounted for them. Then the one who had received five talents approached, and brought him five other talents, And when he had also received two talents, he said, Lord, you have given me two talents; behold, I gained two other talents with them. His master said to him,

"Well done, good and faithful servant." You have been faithful over a little, I will place you over a lot; enter your master's enjoyment. However, he who received a yalenyo and did not multiply it ordered to be punished. — Matthew 25: 19-23

08. WE ARE PERSISTENT

01. PERSISTENTS IN SURVIVAL

No other species on this earth is so stubborn in trying to survive, we never stop wanting to move on, achieve our conquests, achieve our most complex dreams, discover new horizons, go further than we have already come here. Since the creation of the world and the emergence of the first civilizations, man has sought to improve his way of life, increase his knowledge, shape a more evolved existence. From the ancient inhabitants of the caves to date, much has changed due to their persistence.

We persist in not stopping moving forward, not settling for the present, wishing for radical future changes in all sectors of our lives and those of others. If we reach this century, it is because despite the struggles and difficulties we never stopped on the way, from the beginning our race was faced with all kinds of obstacles, but it did not bow its head as a sign of abandonment, on the contrary, it overcame everyone with optimism and showed that it doesn't surrender that easily.

We should be proud of who we are, of everything we have done and have accomplished over the centuries, for never surrendering to our worst fears, countless enemies, plagues, diseases, wars, hunger or misery that may have come to afflict us. We were and will continue to be strong warriors, we will fight hard against whatever it is and we will not give up on our survival. Humanity was created to dominate this planet, nature surrenders to man, we are her masters.

Taking into account what the sacred texts tell us, we read in the book of Genesis: *"And God created man in his image; in the image of God he created him; man and woman created them. And God blessed them, and God said to them, Be fruitful and multiply, and fill the earth, and subdue it; and dominate over the fish of the sea and over the birds of the sky, and over every animal that moves on the earth "*. *Genesis 1: 27,28*

The divine order was that we subjected the land and everything on it under our orders, we are special and sovereign over everything that moves and breathes on this planet, even the birds of the sky and the fish in the oceans owe us obedience and fear.

Therefore, this power that has been given to us elevates humanity to the top of what exists on this planet, we are its greatest glory. The natural insistence we have is what motivates us to grow as people, to evolve, to modernize ourselves day after day. Without this capacity we would not have passed the initial phase of creation and even now we would be living in caves, in the caves in the most complete evolutionary poverty, lost in ignorance and scared by everything around us.

However, the curiosity that is peculiar to us made us discover fire, build the wheel, create means of transport, the metal for the formation of weapons for hunting, fishing and self-defense. We stopped living in dens, tents, shacks and started to build more comfortable houses, beautiful palaces.

We found bronze, silver, gold, we created our first purchase currency, we became traders, we accumulated a lot of wealth. Our species is striding towards a future where it can be seen and remembered as unbeatable, invincible in its purposes. We are the crown of Creation because we dominate the world where we inhabit, because only we have been given such right and authority.

02. PERSISTENTS IN THE EVOLUTION OF OUR KIND

If at present we are a species so developed as to create wings and fly, climb up to outer space in search of other intelligent lives, if through science we research and discover the cure of several evils, if we create more sophisticated armaments every day to to fight our enemies, if we invent and use technology to communicate and keep the whole world in the palm of our hands, it is because we persist in evolving, we do not settle down or stay still in time.

Our species matures year after year, it transforms, undergoes a constant scientific, technological, cultural and social metamorphosis even though these changes do not always bring us benefits, positive results, but even so they help humanity as a whole to evolve, cooperate to that each new generation stays well ahead of the ancient traditions of its ancestors. Today women have more rights and freedoms than before, children, youth and adolescents have their own status that protects them.

Workers with formal contracts have greater labor rights, putting an end to slave labor and without a fair remuneration, employers also benefit from the quality of the workforce. Industries have modernized, new machines and electronic equipment intensify and improve the manufacture of various products, including automobiles. Modernity has transformed the face of many countries and led them to become world leaders in the economy, others follow this advance closely or from afar.

Today's man can no longer even imagine himself living decades ago where nothing he has access to today existed where even a child easily dominates a computer, a cell phone, knows how to surf the internet and make friends on social networks. In the evolved world in which we live in this present age, we are the shining of an era where new values were found.

New ways of thinking and acting were applied to us and this whole apparatus of change makes us look at the past with contempt for the insignificance in which our lives lived. ancestors. However, as we never stop evolving, future generations will also look back and think the same about us. Just as we criticize the use of cell radio by our grandparents, children and grandchildren, they will criticize what we use now.

As technology continues to expand, they will have electronic equipment in their hands so dynamic that they will look at what at this moment is extremely novel with disdain, because for their time what is new today is just a technological backwardness.

But this should not sadden us or wane in hoping for significant changes, agreeing with modernity or making us critical of the idea of evolution. I say this because many are against certain changes because they are longing for times gone by.

We know that for a large number of people there is a deep longing for their childhoods, their youthful times where everything was simpler where interpersonal relationships were more direct, physical contact existed, individuals looked into each other's eyes and clearly heard the sound of your words.

Today, however, this direct relationship has been replaced by electronic devices, it is done through social networks where a friend has ceased to be real to be just a virtual figure emblazoned on a screen. Well, but so what if things have changed and we have reached that point? Imagine how before everything was more difficult, more complicated for us to get in touch with someone who was in another city, state or even in another country? Today we can communicate with a friend or relative who is currently in China or Japan, everything has become so easy! It was worth evolving!

03. PERSISTENTS IN OUR BELIEFS

Since more remote times, the human being has learned to defend his own beliefs in some divine, powerful and sovereign being. It does not matter if it is a solid faith in what the sacred texts teach, as in any god created by human hands like the idols of stick, stone or plaster, the process of belief and worship remains the same today. The truth is that despite all his evolution, man seems to be unable to break free of this ancient tradition, his religiosity is part of his essence, of his deepest self.

"The concept of divinity has assumed, over the centuries, several conceptions, evolving from the most primitive forms from the ancient tribes to the dogmatic definitions of religions. Divinity is, according to those who believe in it, some supernatural, mythological being, with special powers, superior, created spontaneously or by another deity; whose image is often seen as similar to that of man. Worshiped, it is considered as holy, divine or sacred, and / or respected by human beings. Usually, deities are perceived as superior to human beings, controlling or being superior to nature itself.

Deities take a variety of forms, but are often anthropomorphic or zoomorphic. A deity may be masculine, feminine, hermaphroditic or neutral, but it is usually immortal. Sometimes, the deities are identified with elements or phenomena of nature, human virtues or vices or activities inherent to human beings. In addition, it is usual for a particular deity to preside over aspects of man's daily life, such as birth, death, time, fate, etc. Historically, it is not possible to define which was the first tribe to manifest an idea of divinity. The first of these conceptions would have appeared in the Paleolithic and Neolithic periods.

And they would have been manifested by the human feeling of a bond with the Earth and with Nature, cycles and fertility. [1] However, the oldest writings found until today refer to the conceptions coming from the Sumerian, Vedic and Egyptian religions, which arose around 3600 BC.

In order to create explanations for the existence of the elements and beings of nature, as well as to know the meaning of natural phenomena (the storm, the wind, the day and the night, the seasons, etc.), the peoples and tribes of the antiquity conceived several deities that, more often than not, came to have feelings and emotions identical to those of humans.

Hence came the rituals, ceremonies and sacrifices, which were intended to thank the blessings sent by these deities or to appease their wrath, which punished humanity with some calamity. Today, many intellectuals question and speculate about the very concept of Divinity, as well as several historical and archaeological investigations attempt to unravel the events and circumstances in which primitive men created the first religious concepts "

As we can see, each generation creates for itself its own concepts and dogmas about the divinities it chooses to serve and worship and even in the face of all modernity, the need to worship a higher being may remain hidden in your mind as if it somehow depended. exclusively to continue to exist.

Religions exist exactly because of this extreme lack in human life, because of the enormous need to bow to a powerful whole, even without being able to see it. Extreme and uncontrolled religiosity has led to the emergence of several temples, sects, vain doctrines that are used by false pastors, prophets and preachers that aim at easy enrichment.

Who we are? - About the Mysteries of Our Existence

The tithe doctrine is one of the most current ways for these fraudsters to accumulate alarming wealth, I can mention here the leader of the Universal Church of the Kingdom of God, Edir Macedo, as the brain that triggered the emergence of neo-Pentecostal religions in Brazil whose doctrinal basis called "The Doctrine of Prosperity "has led thousands of people to dispose of their last savings and material goods to donate everything in favor of the false belief that divine blessings can be bought.

The Holy Scriptures make it very clear that God does not negotiate his blessings, everything he does for his children is out of pure love and mercy. Christ, in one of his many preachings, warned:

"Ask, and it will be given to you; seek, and you will find; knock, and the door will be opened to you. For everyone who asks receives; what seeks, finds; and to the knocker, the door will be opened. Who of you, if your son asks for bread, will give you a stone? Or, if you ask for fish will you give him a snake? If you, despite being bad, know how to give good things to your children, how much more will your Father who is in heaven give good things to those who ask him! " Matthew 7: 7-11

We observe, through the explanation of Christ, that it is enough to ask for something that we need from our Heavenly Father and we will be attended to, in no time the biblical text mentions that we need to give something in return in order to receive what we need. Therefore, the doctrine of prosperity created by Macedo and taught in the temples of his church is false and abusive, aiming only to engage and extort the faithful who attend there. But this occurs almost all over the world because of the immense religious ignorance of those who refuse to study the Scriptures in order to believe in the teachings of people like this who do not serve the Kingdom of God, but their own interests.

"In the 19th century, a theological trend arose in the United States whose central axis was the commercialization of the Christian faith based on the distortion of biblical teachings. As a differential, he fervently defended the accumulation of material wealth on earth.

To achieve their goals, those responsible for the dissemination of the ideals of Prosperity Theology, seek to gather as many followers as possible, since by "prosperous enterprise" they acquire greater contributions by their faithful in their temples ".

It is not for nothing that Edir Macedo is today considered the richest bishop in the world, his wealth is enormous and he is exalted because of this, claiming that God honored him for his faith. But when we read what Jesus said to his disciples about the material wealth accumulated in this world, we see a contrast in Macedo's doctrine, because while he affirms that the children of God must boast great treasures on earth, the Savior of humanity guides us to accumulate treasures in the heaven (Or in the spiritual life)

"Do not set up treasures on the land, where moth and rust consume everything, and where thieves mine and steal. But gather up treasures in the sky, where neither moth nor rust consume, and where thieves neither mine nor steal. Because wherever your treasure is, there will also be your heart. Matthew 6: 19-21

And in another part of Scripture, we read:

"Because the love of money is the root of all kinds of evils; and in that greed some have strayed from the faith, and pierced themselves with many pains ". 1 Timothy 6:10

Therefore, what can be seen is that Edir Macedo is opposed to Christ. This man's disbelief in the teachings of the Son of God is so visible that he dares to disagree with the Gospel...

Who we are? - About the Mysteries of Our Existence

And started teaching everything contrary to what the Holy Book tells us. Jesus Christ taught humility, Macedo, a delighted life filled with enormous riches. The Master made it clear that the Father does not negotiate his blessings, Bishop Edir Macedo guarantees that if someone sells everything he has and donates to his temples, God will return everything in return as he needs some material good that we have. Let us note there a huge contrast in the bishop's teaching in relation to what the Gospels tell us, because while Jesus sent that rich young man to sell all his goods and donate to the poorest the doctrine of prosperity is based on everything being given to the founder of Universal Church.

"And Jesus, looking at him, loved him and said to him: You lack one thing: go, sell everything you have, and give it to the poor, and you will have a treasure in heaven; and come, take the cross, and follow me. But he, sorry for this word, withdrew sadly; because he had many properties. Then Jesus, looking around, said to his disciples: How difficult it is for those who have wealth to enter the kingdom of God!" Mark 10: 21-23

That man was always attached to his material wealth and did not have the strength to give up everything he owned. He owned many properties, he was powerful in those lands with many employees at his service. The idea of giving up all his assets to follow a prophet, even though seeing him perform so many miracles, discouraged him. In the same way it happens today with all those who love money and its goods more than God.

They place their current position above the spiritual life that we will certainly have after death, because we are a living soul living in physical bodies that gradually grow old, they get sick and die. Edir Macedo prospered financially due to the ignorance of people who stick to the faith in a false doctrine. The need of most human beings to seek a religious ideology that guarantees or at least promises prosperity in this world.

The cure of their inner or physical ills, leads them to commit insane acts such as giving to these mercenaries what little they have left. . But that will never change, there will always be those who preach the lie and those who give credit to their fables. When writing a letter to his disciple Timothy more than two thousand years ago, Paul warned:

"The Spirit clearly says that in recent times some will abandon the faith and follow deceiving spirits and doctrines of demons. Such teachings come from hypocritical and lying men, who have a seared conscience and forbid the marriage and consumption of food that God created to be received with thanksgiving by those who believe and know the truth. For everything that God created is good, and nothing should be rejected if it is received with thanksgiving, for it is sanctified by the word of God and by prayer. If you pass these instructions on to your brothers, you will be a good minister of Christ Jesus, nourished by the truths of faith and the good doctrine that you have followed. Reject, however, the profane fables of old women and exercise piety ". 1 Timothy 4:

FINAL CONSIDERATIONS

This work aims to awaken in my readers the importance of understanding the real reasons that lead them to be here, on this planet, as well as encouraging them to try to discover their true missions as members of a greater purpose in cultural, scientific and cultural growth. technological development of humanity. We cannot believe or accept the idea that we were born, grew up and died by chance, without a specific purpose, in a random way and without an objective beyond what we can see or imagine. We are part of a project, a plan devised by a superior force that tends to make the human race reach the peak of its evolution.

BIBLIOGRAPHY

https://www.educamaisbrasil.com.br/enem/biologia/teoria-da-evolucao

Wagner Dias Ferreira - Lawyer and member of the OAB / MG Human Rights Commission

https://www.bbc.com/portuguese/geral-56587394

https://veja.abril.com.br/ciencia/enviar-o-homem-a-marte-ate-2030-e-prioridade-da-nasa/

https://www.ipea.gov.br/cts/pt/central-de-conteudo/artigos/artigos/116-a-ciencia-e-a-tecnologia-como-estrategia-de-desenvolvimento

http://ve.scielo.org/scielo.php?script=sci_arttext&pid=S0378-18442006000700003

https://www.boog.com.br/artigos/atitudes-que-fazem-a-diferenca

https://en.wikipedia.org/wiki/Livre-arb%C3%ADtrio

https://agazETAacre.com/2014/05/artigos/contradicoes-da-vida-real/

https://en.wikipedia.org/wiki/Dclaim

https://www.cartacapital.com.br/blogs/dialogos-da-fe/teologia-da-prosperidade-o-mercado-da-fe-e-a-fe-mercadologica/

Lightning Source UK Ltd.
Milton Keynes UK
UKHW021256180621
385747UK00002B/382